Glossary of
Soil Science Terms

1997

SOIL SCIENCE SOCIETY OF AMERICA
677 South Segoe Road • Madison WI • 53711
phone: (608) 273-8080
home page: www.soils.org

Soil Science Society of America, Inc.
677 South Segoe Road, Madison, Wisconsin 53711 USA

Library of Congress Cataloging-in-Publication Data

Library of Congress Catalog Card Number: 96-72488

Printed in the United States of America

CONTENTS

INTRODUCTION

This February 1997 revision of the *Glossary of Soil Science Terms* replaces the July 1987 edition and includes major revisions and additions. A *Glossary* committee was appointed in May 1993 to revise the 1987 edition. Members of the committee were:

Charles B. Roth, Div. S-9, Chair   George F. Hall, Div. S-5
Alan R. Mitchell, Div. S-1   Robert E. Sojka, Div. S-6
Paul R. Bloom, Div. S-2   Russell D. Briggs, Div. S-7
Mark S. Coyne, Div. S-3   Randy J. Killorn, Div. S-8
Dale T. Westerman, Div. S-4   Paula Gale replaced K.R. Reddy, Div. S-10
Jerry M. Bigham replaced Robert J. Luxmoore, SSSA Editor-in-Chief, *ex officio*
Lloyd R. Hossner replaced D. Keith Cassel, SSSAJ Editor, *ex officio*

Work by this committee resulted in the revision of approximately two-thirds of the 1514 terms contained in the July 1987 edition of the *Glossary*. The "obsolete" list was moved into the main body of the glossary to facilitate the finding of these terms with addition of "no longer used in SSSA publications" or "Not used in current U.S. system of soil taxonomy" where appropriate. A total of 113 terms were deleted from the July 1987 edition and 420 new terms were added to this revision of the Glossary. Three of the nine tables contained in the July 1987 revision were deleted while three more were revised radically. The other three tables had only slight modifications and two new tables were added to this revision of the *Glossary*.

The SSSA has published definitions or glossaries of soil science terms since 1956. The first five published lists appeared in the *SSSA Proceedings* from 1956 through 1965. These glossaries were developed by various terminology committees of SSSA. From 1968 through 1984 revisions in terminology were handled by SSSA division terminology committees. New terms were carried as appendices in the five revised editions of the Glossary published from 1971 through 1979, plus a supplement in October 1982. The 1984 revision incorporated a policy that all acceptable terms appear in a single alphabetical list. Appendices would include obsolete terms, tabular information, proposed new terms or guidelines, etc. This revision of the *Glossary* does not have a separate list of obsolete terms.

Measurements included with terms are in SI units to conform with SSSA policy requiring SI units for all publications. Conversion factors for SI and non-SI units are included at the end of this *Glossary*.

None of the terms in the *Glossary of Soil Science Terms* are considered official by the SSSA. They are published in an effort to provide a foundation for common understanding in communications covering soil science. Suggestions for revisions in the *Glossary* should be sent to the S374 Glossary of Soil Science Terms Committee, c/o SSSA Headquarters Office, 677 South Segoe Road, Madison, WI 53711 or by e-mail to "gloss_soil@agronomy.org".

Sincere thanks are expressed to the many members of the society who have aided in the development of this glossary over the years.

Glossary of Soil Science Terms

A

A horizon–See **soil horizon** and **Appendix II**.

abiontic enzymes–Enzymes (exclusive of live cells) that are (i) excreted by live cells during growth and division; (ii) attached to cell debris and dead cells; (iii) leaked into soil solution from extant or lyzed cells but whose original functional location was on or within the cell. Synonymous with exoenzymes.

abiotic factor–A physical, meteorological, geological, or chemical aspect of the environment.

ablation till–A general term for loose, relatively permeable material, either contained within or accumulated on the surface of a glacier deposited during the downwasting of nearly static glacial ice.

absorptance–The ratio of the radiant flux absorbed by a body to that incident upon it. Also called absorption factor.

absorption–Uptake of matter or energy by a substance.

absorption, active–Movement of ions and water into the plant root because of metabolic processes by the root, frequently against an electrochemical potential gradient.

absorption, passive–Movement of ions and water into the plant root from diffusion along a chemical potential gradient.

accelerated erosion–See **erosion**, *accelerated erosion*

access tube–Small diameter tube (typically about 50 millimeters) inserted through the soil root zone to provide passage of a **neutron probe** to determine the water content of soil at various depths.

acetylene-block assay–A technique for demonstrating or estimating denitrification by measuring nitrous oxide (N_2O) released from acetylene-treated soil. Acetylene inhibits nitrous oxide reduction to dinitrogen (N_2) by denitrifying bacteria.

acetylene-reduction assay–A technique for demonstrating or estimating nitrogenase activity by measuring the rate of acetylene (C_2H_2) reduction to ethylene (C_2H_4).

acid precipitation–Atmospheric precipitation that is below pH 7 and is often composed of the hydrolyzed by-products from oxidized halogen, nitrogen, and sulfur substances.

acid soil–Soil with a pH value <7.0.

acidic cations–Cations that, on being added to water, undergo hydrolysis resulting in an acidic solution. Hydrated acidic cations donate protons to water to form hydronium ions (H_3O^+) and thus in aqueous solutions are acids according to the definition given by Bronsted. Examples in soils are Al^{3+} and Fe^{3+}.

acidity, active–(no longer used in SSSA publications) The activity of hydrogen ion in the aqueous phase of a soil expressed as a pH value.

acidity, exchange–(no longer used in SSSA publications) The acidity of a soil that can be neutralized by lime or a solution buffered in the range of 7 to 8. See also **acidity, total**.

acidity, exchangeable–See **acidity, salt-replaceable.**

acidity, free–(no longer used in SSSA publications) The titratable acidity in the aqueous phase of a soil.

acidity, reserve–See **acidity, residual**.

acidity, residual–Soil acidity that is neutralized by lime or a buffered salt solution to raise the pH to a specified value (usually 7.0 or 8.0) but which cannot be replaced by an unbuffered salt solution. It can be calculated by subtraction of salt replaceable acidity from total acidity. See also **acidity, salt-replaceable** and **acidity, total**.

acidity, salt-replaceable–The aluminum and hydrogen that can be replaced from an acid soil by an unbuffered salt solution such as KCl or NaCl.

acidity, total–The total acidity including **residual** and **exchangeable acidity**. Often it is calculated by subtraction of exchangeable bases from the cation exchange capacity determined by ammonium exchange at pH 7.0. It can be determined directly using pH buffer-salt mixtures (e.g. $BaCl_2$ plus triethanolamine, pH 8.0 or 8.2) and titrating the basicity neutralized after reaction with a soil.

acidulation–The process of treating a fertilizer source with an acid. The most common process is treatment of phosphate rock with an acid (or mixture of acids) such as sulfuric, nitric, or phosphoric acid.

activation energy–A term used in kinetics to indicate the amount of energy required to bring all molecules in one mole of a substance to their reactive state at a given temperature. Conceptually, this energy barrier must be overcome to get a reaction to go forward. At higher activation energies, reactions are slower if temperature and composition are constant. It is usually determined from an Arrhenius plot of the inverse of the absolute temperature vs. rates of reaction at different temperatures.

active layer–The top layer of ground subject to annual thawing and freezing in areas underlain by permafrost.

activity (chemical)–(i) A dimensionless measure of the deviation of the chemical potential of a substance from its value at a standard state. It is defined by the equation: $\mu = \mu° + RT \ln a$, where μ is the chemical potential at activity $= a$, $\mu°$ is the chemical potential in the standard state (where $a = 1.0$), R is the molar gas constant, and T is the absolute temperature. In solution $a =$ molal concentration at infinite dilution (molal concentration = molar concentration at low concentrations), and in gases $a =$ partial pressure in atmospheres.
(ii) Informally, in solution, it may be taken as the effective concentration of a substance. See also **activity coefficient**.

activity coefficient–The ratio between the **activity (chemical)** and the concentration of a substance in solution. Activity of component n is usually indicated by (n) and concentration by [n].

adenylate energy charge ratio (EC)–A measure of the metabolic and growth state of microorganisms and microbial communities. The energy charge ratio is calculated using the formula: $EC = (ATP + \frac{1}{2}ADP)/(ATP + ADP + AMP)$. The denominator represents the total adenylate pool; the numerator, the portion charged with high energy phosphate bonds.

adsorption–The process by which atoms, molecules, or ions are taken up from the soil solution or soil atmosphere and retained on the surfaces of solids by chemical or physical binding.

adsorption complex–Collection of various organic and inorganic substances in soil that are capable of adsorbing ions and molecules.

adsorption isotherm–A graph of the quantity of a given chemical species bound to an adsorption complex, at fixed temperature, as a function of the concentration of the species in a solution that is in equilibrium with the complex. Called an isotherm only because adsorption experiments are done at constant temperature.

advance time–See **irrigation**, *advance time.*

advection–See **convection**.

aerate–To allow or promote exchange of soil gases with atmospheric gases.

aeration porosity–See **air porosity**.

aeration, soil–The process by which air in the soil is replaced by air from the atmosphere. In a well-aerated soil, the soil air is very similar in composition to the atmosphere above the soil. Poorly aerated soils usually contain a much higher content of CO_2 and a lower content of O_2 than the atmosphere above the soil. The rate of aeration depends largely on the volume and continuity of air-filled pores within the soil.

aerobic–(i) Having molecular oxygen as a part of the environment. (ii) Growing only in the presence of molecular oxygen, such as aerobic organisms. (iii) Occurring only in the presence of molecular oxygen (said of chemical or biochemical processes such as aerobic decomposition).

aerobic digestion–The partial biological decomposition of suspended organic matter in waste water or sewage in aerated conditions.

aggregate–A group of primary soil particles that cohere to each other more strongly than to other surrounding particles.

aggregation–The process whereby primary soil particles (sand, silt, clay) are bound together, usually by natural forces and substances derived from root exudates and microbial activity.

agric horizon–A mineral soil horizon in which clay, silt and humus derived from an overlying cultivated and fertilized layer have accumulated. The wormholes and illuvial clay, silt and humus, occupy at least 5% of the horizon by volume. The illuvial clay and humus occur as horizontal lamellae or fibers, or as coatings on ped surfaces or in wormholes.

agrichemicals–Chemical materials used in agriculture.

agroforestry–Any type of multiple cropping land-use that entails complementary relations between tree and agricultural crops and produces some combination of food, fruit, fodder, fuel, wood, mulches, or other products.

agrohydrology–See **hydrology.**

agronomic rate–The rate at which fertilizers, organic wastes or other amendments can be added to soils for optimum plant growth.

agronomy–The theory and practice of crop production and soil management.

air dry–(i) The state of dryness at equilibrium with the water content in the surrounding atmosphere. The actual water content will depend upon the relative humidity and temperature of the surrounding atmosphere. (ii) To allow to reach equilibrium in water content with the surrounding atmosphere.

air entry value–The value of water content or potential at which air first enters a porous media.

air (-filled) porosity–The fraction of the bulk volume of soil that is filled with air at any given time or under a given condition, such as a specified soil-water content or soil-water matric potential.

alban–A cutan that is light-colored in thin section because of the reduction and translocation of iron.

albedo–The ratio of the amount of solar radiation reflected by a body to the amount incident upon it, often expressed as a percentage, as, the albedo of the earth is 34%.

albic horizon–A mineral soil horizon from which clay and free iron oxides have been removed or in which the oxides have been segregated to the extent that the color of the horizon is determined primarily by the color of the primary sand and silt particles rather than by coatings on these particles.

Albolls–**Mollisols** that have an **albic horizon** immediately below the **mollic epipedon**. These soils have an **argillic** or **natric horizon** and mottles, iron-manganese **concretions**, or both, within the **albic**, **argillic** or **natric** horizon. (A suborder in the U.S. system of soil taxonomy.)

Alfisols–Mineral soils that have **umbric** or **ochric epipedons**, **argillic horizons**, and that hold water at <1.5 MPa tension during at least 90 days when the soil is warm enough for plants to grow outdoors. Alfisols have a mean annual soil temperature of <8° C or a base saturation in the lower part of the **argillic horizon** of 35% or more when measured at pH 8.2. (An order in the U.S. system of soil taxonomy.)

alkali soil–(no longer used in SSSA publications) (i) A soil with a pH of 8.5 or higher or with a exchangeable sodium ratio greater than 0.15. (ii) A soil that contains sufficient sodium to interfere with the growth of most crop plants. See also **saline-sodic soil** and **sodic soil**.

alkaline soil–Soil with a pH value >7.0.

alkalinity, soil–The degree or intensity of alkalinity in a soil, expressed by a value >7.0 for the soil pH.

allelopathy–See **antagonism**.

allochthonous flora–Organisms that are not indigenous to the soil but that enter in precipitation, diseased tissues, manure, sewage, etc. They may persist but do not notably contribute to ecologically significant transformations or interactions.

allophane–An aluminosilicate with primarily short-range structural order. Occurs as exceedingly small spherical particles especially in soils formed from volcanic ash.

alluvial–Pertaining to processes or materials associated with transportation or deposition by running water.

Alluvial soil–(i) A soil developing from recently deposited alluvium and exhibiting essentially no horizon development or modification of the recently deposited materials. (ii) When capitalized the term refers to a great soil group of the azonal order consisting of soils with little or no modification of the recent sediment in which they are forming. (Not used in current U.S. system of soil taxonomy.)

alluvium–Sediments deposited by running water of streams and rivers. It may occur on terraces well above present streams, on the present flood plains or deltas, or as a fan at the base of a slope.

Alpine Meadow soils–A great soil group of the intrazonal order, comprised of dark soils of grassy meadows at altitudes above the timberline. (Not used in current U.S. system of soil taxonomy.)

amensalism–An interaction between two organisms in which one organism is suppressed by the other (such as suppression of one organism by toxins produced by the second).

ammonia volatilization–Mass transfer of nitrogen as ammonia gas from soil, plant, or liquid systems to the atmosphere.

ammoniation–The process of introducing various ammonium sources into other fertilizer sources forming ammoniated compounds. Ammonium polyphosphates and ammoniated superphosphate are ammoniated compounds.

ammonification–The biological process leading to ammoniacal nitrogen formation from nitrogen-containing organic compounds.

ammonium fixation–The process of entrapment of ammonium ions in interlayer spaces of phyllosilicates, in sites similar to K^+ in micas. Smectites, illites and vermiculites all can fix ammonium, but vermiculite has the greatest capacity. The fixation may occur spontaneously in aqueous suspensions, or as a result of heating to remove interlayer water. Ammonium ions in collapsed interlayer spaces are exchangeable only after expansion of the interlayer. See also **potassium fixation**.

ammonium phosphate–A generic class of compounds used as phosphorus fertilizers. Manufactured by the reaction of anhydrous ammonia with orthophosphoric acid or superphosphoric acid to produce either solid or liquid products.

amorphous material–Noncrystalline constituents that either do not fit the definition of allophane or it is not certain if the constituent meets allophane criteria.

anaerobic–(i) The absence of molecular oxygen. (ii) Growing in the absence of molecular oxygen (such as anaerobic bacteria). (iii) Occurring in the absence of molecular oxygen (as a biochemical process).

anaerobic respiration–The metabolic process whereby electrons are transferred from a reduced compound (usually organic) to an inorganic acceptor molecule other than oxygen. The most common acceptors are carbonate, sulfate, and nitrate. See also **denitrification**.

anchor–See **tillage**, *anchor*.

Andepts–Previous to 1994 this term was used to indicate Inceptisols that have formed either in vitric pyroclastic materials, or have low bulk density and large amounts of amorphous materials, or both. The term was dropped as a suborder in the 1994 revision of the USDA, *Keys to soil taxonomy*.

andic–Soil properties related to volcanic origin of materials. The properties include organic carbon content, bulk density, phosphate retention, and iron and aluminum extractable with ammonium oxalate.

Andisols–Mineral soils that are dominated by andic soil properties in 60 percent or more of their thickness. (An order in the U.S. system of soil taxonomy.)

angle of repose–The maximum angle of slope (measured from a horizontal plane) at which loose, cohesionless material will come to rest.

Aquults–**Ultisols** that are saturated with water for periods long enough to limit their use for most crops other than pasture or woodland unless they are artificially drained. **Aquults** have mottles, iron-manganese concretions or gray colors immediately below the A1 or Ap horizons and gray colors in the argillic horizon. (A suborder in the U.S. system of soil taxonomy.)

arable land–Land so located that production of cultivated crops is economical and practical.

arbuscule–Specialized dendritic (highly branched) structure formed within root cortical cells by **endomycorrhizal** fungus. See also **vesicular arbuscular**.

archaebacteria–(i)Prokaryotes with cell walls that lack murein, having ether bonds in their membrane phospholipids, that are characterized by growth in extreme environments. (ii) A primary biological kingdom distinct from both **eubacteria** and **eukaryotes**.

Arents–**Entisols** that contains recognizable fragments of pedogenic horizons that have been mixed by mechanical disturbance. **Arents** are not saturated with water for periods long enough to limit their use for most crops. (A suborder in the U.S. system of soil taxonomy.)

Argids–**Aridisols** that have an **argillic** or a **natric horizon**. (A suborder in the U.S. system of soil taxonomy.)

argillan–A **cutan** composed dominantly of clay minerals.

argillic horizon–A mineral soil horizon that is characterized by the illuvial accumulation of **phyllosilicate** clays. The **argillic horizon** has a certain minimum thickness depending on the thickness of the solum, a minimum quantity of clay in comparison with an overlying eluvial horizon depending on the clay content of the eluvial horizon, and usually has coatings of oriented clay on the surface of pores or peds or bridging sand grains.

aridic–A soil moisture regime that has no water available for plants for more than half the cumulative time that the soil temperature at 50 cm below the surface is >5° C, and has no period as long as 90 consecutive days when there is water for plants while the soil temperature at 50 cm is continuously >8° C.

Aridisols–Mineral soils that have an **aridic** moisture regime, an **ochric epipedon**, and other pedogenic horizons but no **oxic horizon**. (An order in the U.S. system of soil taxonomy.)

artificial manure–(no longer used in SSSA publications) In European usage denotes commercial fertilizers.

aseptic–Free from pathogenic or contaminating organisms.

ash (volcanic)–Unconsolidated, pyroclastic material less than 2 mm in all dimensions. Commonly called "volcanic ash". Compare **cinders**, **lapilli**, **tephra**.

aspect–The direction toward which a slope faces with respect to the compass or to the rays of the sun.

assimilation–The incorporation of inorganic or organic substances into cell constituents.

association, soil–See **soil association**.

associative dinitrogen fixation–A close interaction between a free-living diazotrophic organism and a higher plant that results in enhanced dinitrogen fixation rates.

associative symbiosis–A close but relatively casual interaction between two dissimilar organisms or biological systems. The association may be mutually beneficial but is not required to accomplish specific functions. See also **commensalism, symbiosis.**

attapulgite clay–See **palygorskite.**

Atterberg limits–The collective designation of seven so-called limits of consistency of fine-grained soils, suggested by Albert Atterberg, 1911-1912, but with current usage usually retaining only the liquid limit, the plastic limit, and the plasticity number (or index). See also **consistency, liquid limit, plastic limit,** and **plasticity number.**

autochthonous–Microorganisms and/or substances indigenous to a given ecosystem; the true inhabitants of an ecosystem; referring to the common microbiota of the body of soil microorganisms that tend to remain constant despite fluctuations in the quantity of fermentable organic matter.

autochthonous flora–(i) That portion of the micro flora presumed to subsist on the more resistant soil organic matter and little affected by the addition of fresh organic materials. (ii) Microorganisms indigenous to a given ecosystem; the true inhabitants of an ecosystem; referring to the common microbiota of the body of soil microorganisms that tend to remain constant despite constant fluctuations in the quantity of fermentable organic matter. Contrast with **zymogenous flora.** Also termed **oligotrophs.**

autotroph–An organism capable of utilizing CO_2 or carbonates as a sole source of carbon and obtaining energy for carbon reduction and biosynthetic processes from radiant energy (**photoautotroph** or **photolithotroph**) or oxidation of inorganic substances (**chemoautotroph** or **chemolithotroph**).

autotrophic nitrification–Oxidation of ammonium to nitrate through the combined action of two **chemoautotrophic** bacteria, one forming nitrite from ammonium and the other oxidizing nitrite to nitrate.

available nutrients–(i) The amount of soil nutrient in chemical forms accessible to plant roots or compounds likely to be convertible to such forms during the growing season. and (ii) The contents of legally designated "available" nutrients in fertilizers determined by specified laboratory procedures which in most states constitute the legal basis for guarantees.

available water (capacity)–The amount of water released between in situ field capacity and the permanent wilting point (usually estimated by water content at soil matric potential of -1.5 MPa). It is not the portion of water that can be absorbed by plant roots, which is plant specific. See also **nonlimiting water range.**

avalanche–A large mass of snow, ice, soil, or rock, or mixtures of these materials, falling, sliding, or flowing very rapidly under the force of gravity. Velocities may sometimes exceed 500 km/hr.

azonal soils–Soils without distinct genetic horizons. (Not used in current U.S. system of soil taxonomy.)

B

B horizon–See **soil horizon** and **Appendix II.**

backfurrow–See **tillage,** *backfurrow.*

backslope–The hillslope position that forms the steepest, and generally linear, middle portion of the slope. In profile, backslopes are bounded by a convex shoulder above and a concave footslope below.

backswamp–A flood-plain landform. Extensive, marshy, or swampy, depressed areas of flood plains between natural levees and valley sides or terraces.

bacteroid–An altered form of bacterial cells. Refers particularly to the swollen, irregular vacuolated cells of *Rhizobium* and *Bradyrhizobium* in legume nodules.

badland–In Soil Survey a map-unit that is a type of miscellaneous area, which is generally devoid of vegetation , is intricately dissected by a fine, drainage network with a high drainage density and has short, steep slopes with narrow interfluves resulting from erosion of soft geologic materials. Most common in arid or semiarid regions. See also **miscellaneous area**.

band application–See **banding**.

banding–A method of fertilizer or other agrichemical application above, below, or alongside the planted seed row. Refers to either placement of fertilizers close to the seed at planting or subsurface applications of solids or fluids in strips before or after planting. Also referred to as **band application**.

bar–(i) A generic term for ridge-like accumulations of sand , gravel, or other unconsolidated material formed in the channel, along the banks, or at the mouth of a streams or formed by waves or currents as offshore features in large lakes or oceans. (ii) A unit of pressure equal to one million dynes per square centimeter. Megapasal is the preferred unit for pressure in SSSA publications.

basal till–Unconsolidated material deposited and compacted beneath a glacier and having a relatively high bulk density. See also **till**, **ablation till**, **lodgement till**.

base level–The theoretical limit or lowest level toward which erosion of the Earth's surface constantly progresses but seldom, if ever, reaches; especially the level below which a stream cannot erode its bed. The general or ultimate base level for the land surface is sea level, but temporary base levels may exist locally.

base saturation–The ratio of the quantity of exchangeable bases to the cation exchange capacity. The value of the base saturation varies according to whether the cation exchange capacity includes only the salt extractable acidity (see **cation exchange capacity**) or the total acidity determined at pH 7 or 8. Often expressed as a percent.

basic fertilizer–One that, after application to and reaction with soil, decreases residual acidity and increases soil pH.

basic slag–A by-product in the manufacture of steel, containing lime, phosphorus, and small amounts of other plant nutrients such as sulfur, manganese, and iron.

batch culture–A method for culturing organisms in which the organism and supporting nutritive medium are added to a closed system. Contrast with **chemostat**.

bay– (i) Any terrestrial formation resembling a bay of the sea, as a recess or extension of lowland along a river valley or within a curve in a range of hills, or an arm of a prairie extending into, or partly surrounded by, a forest. (ii) A **Carolina Bay**.

beach–A gently sloping area adjacent to a lake or ocean that lies between the low and high water marks, which is devoid of vegetation, and is composed of unconsolidated material, typically sand or gravel, deposited by waves or tides.

bed–(i) Geologic - The smallest, formal lithostratigraphic unit of sedimentary rocks. The designation of a bed or a unit of beds as a formally named lithostratigraphic unit generally should be limited to certain distinctive beds whose recognition is particularly useful. Coal beds, oil sands, and other layers of economic importance commonly are named, but such units and their names usually are not a part of formal stratigraphic nomenclature. (ii) Agronomic - A raised (usually) cultivated area between furrows or wheel tracks of tractors specially prepared, managed and/or irrigated to promote the production of a planted crop.

bed load–See **erosion,** *bed load.*

bed planting–See **tillage,** *bed planting.*

bed shaper–See **tillage,** *bed shaper.*

bedding–See **tillage,** *bedding.*

bedrock–A general term for the solid rock that underlies the soil and other unconsolidated material or that is exposed at the surface.

bentonite–A relatively soft rock formed by chemical alteration of glassy, high silica content volcanic ash. This material shows extensive swelling in water and has a high **specific surface area**. The principal mineral constituent is clay size smectite.

bioassay–A method for quantitatively measuring a substance by its effect on the growth of a suitable microorganism, plant, or animal under controlled conditions.

biodegradable– A substance able to be decomposed by biological processes.

biological availability–That portion of a chemical compound or element that can be taken up readily by living organisms.

biological denitrification–See **denitrification**.

biological immobilization–See **immobilization** and **biological interchange**.

biological interchange–The interchange of elements between organic and inorganic states in a soil or other substrate through the action of living organisms. It results from the biological decomposition of organic compounds with the liberation of inorganic materials (mineralization) and the utilization of inorganic materials with synthesis of microbial tissue (immobilization).

biomass–(i) The total mass of living organisms in a given volume or mass of soil. (ii) The total weight of all organisms in a particular environment. See also **microbial biomass**.

bioremediation–The use of biological agents to reclaim soil and water polluted by substances hazardous to the environment or human health.

biosequence–A group of related soils that differ, one from the other, primarily because of differences in kinds and numbers of plants and soil organisms as a soil-forming factor.

biotic enzymes–Enzymes associated with viable proliferating cells located (i) intracellularly in cell protoplasm; (ii) in the periplasmic space; (iii) at the outer cell surfaces.

biotite–A brown, trioctahedral layer silicate of the mica group with Fe(II) and Mg in the octahedral layer and Si and Al in a ratio of 3:1 in the tetrahedral layer. See also **Appendix I, Table A3**.

birefringence–The numerical difference between the highest and lowest refractive index of a mineral. Minerals with birefringence exhibit interference colors in thin section when viewed with crossed-polarized light.

birnessite – $(Na_{0.7}Ca_{0.3})Mn_7O_{14}\cdot2.8H_2O$ A black manganese oxide that is common in iron-manganese nodules of soils. It has a layer structure.

bisect–A profile of plants and soil showing the vertical and lateral distribution of roots and tops in their natural position.

bisequal–Soils in which two sequa have formed, one above the other, in the same deposit.

biuret–$H_2NCONHCONH_2$ A product formed at high temperature during the manufacturing of urea. It is toxic to plants. Also called **carbamoylurea**.

Black Earth–A term used by some as synonymous with "**Chernozem**;" by others (in Australia) to describe self-mulching black clays. (Not used in current U.S. system of soil taxonomy.)

Black Soils–A term used in Canada to describe soils with dark-colored surface horizons of the black (**Chernozem**) zone; includes **Black Earth** or **Chernozem**, Wiesenboden, **Solonetz**, etc. (Not used in current U.S. system of soil taxonomy.)

bleicherde–The light-colored, leached A2 (E) horizon of **Podzol** soils.

block checking–See **tillage**, *block*.

block thinning–See **tillage**, *block*.

blocky soil structure–A shape of soil structure. See also **soil structure** and **soil structure shapes**.

blowout–A hollow or depression of the land surface, which is generally saucer or trough-shaped, formed by wind erosion especially in an area of shifting sand, loose soil, or where vegetation is disturbed or destroyed. See also **miscellaneous area**.

blown-out land–In soil survey a map-unit which is a type of **miscellaneous area** from which most of the soil has been removed by wind erosion. The areas are generally shallow depressions with flat, irregular floors, which in some instances have a layer of pebbles or cobbles.

BOD (biochemical oxygen demand)–The quantity of oxygen used in the biochemical oxidation of organic and inorganic matter in a specified time, at a specified temperature, and in specified conditions. An indirect measure of the concentration of biologically degradable material present in organic wastes.

bog–A peat-accumulating wetland that has no significant inflows or outflows and supports acidophilic mosses, particularly *Sphagnum*. See also **fen, marshes, pocosin, swamp**, and **wetland**.

bog iron ore–Impure ferruginous deposits developed in **bogs** or **swamps** by the chemical or biochemical oxidation of iron carried in solution.

Bog soil–A great soil group of the intrazonal order and hydromorphic suborder. Includes **muck** and **peat**. (Not used in current U.S. system of soil taxonomy.)

boom–See **irrigation**, *sprinkler irrigation systems, boom.*

boom center pivot–See **irrigation**, *sprinkler irrigation systems, boom center pivot.*

boom corner pivot–See **irrigation**, *sprinkler irrigation systems, boom corner pivot.*

boom lateral move–See **irrigation**, *sprinkler irrigation systems, boom lateral move.*

boom microirrigation–See **irrigation**, *sprinkler irrigation systems, boom microirrigation.*

boom mist irrigation–See **irrigation**, *sprinkler irrigation systems, boom mist irrigation.*

boom nozzle –See **irrigation**, *sprinkler irrigation systems, boom nozzle.*

boom side-move sprinkler–See **irrigation**, *sprinkler irrigation systems, boom side-move sprinkler.*

boom side-roll sprinkler–See **irrigation**, *sprinkler irrigation systems, boom side-roll sprinkler.*

boom sprinkler distribution pattern–See **irrigation**, *sprinkler irrigation systems, boom sprinkler distribution pattern.*

boom towed sprinkler–See **irrigation**, *sprinkler irrigation systems, boom towed sprinkler.*

Boralfs–**Alfisols** that have formed in cool places. **Boralfs** have **frigid** or **cryic** but not **pergelic** temperature regimes, and have **udic** moisture regimes. **Boralfs** are not saturated with water for periods long enough to limit their use for most crops. (A suborder in the U.S. system of soil taxonomy.)

border dikes–See **irrigation**, *border dikes.*

border ditch–See **irrigation**, *border ditch.*

border-strip–See **irrigation**, *border-strip.*

Borolls–**Mollisols** with a mean annual soil temperature of <8° C that are never dry for 60 consecutive days or more within the 90 days following the summer solstice. **Borolls** do not contain material that has a $CaCO_3$ equivalent >400 g kg^{-1} unless they have a calcic horizon, and they are not saturated with water for periods long enough to limit their use for most crops. (Λ suborder in the U.S. system of soil taxonomy.)

bottomland–See **flood plain**.

boulders–Rock or mineral fragments >600 mm in diameter. See also **rock fragments**.

bouldery–Containing appreciable quantities of boulders. See also **rock fragments**.

bradyrhizobia–Collective common name for the genus *Bradyrhizobium*. See also **rhizobia.**

braided stream–A channel or stream with multiple channels that interweave as a result of repeated bifurcation and convergence of flow around interchannel bars, resembling (in plan view) the strands of a complex braid. Braiding is generally confined to broad, shallow streams of low sinuosity, high bedload, non-cohesive bank material, and a steep gradient.

breakthrough curve–The relative solute concentration in the outflow from a column of soil or porous medium after a step change in solute concentration has been applied to the inlet end of the column, plotted against the volume of outflow (often in number of pore volumes).

breccia–A coarse grained, clastic rock composed of angular fragments (>2mm) bonded by a mineral cement or in a finer-grained matrix of varying composition and origin.

capillary porosity–(no longer used in SSSA publications) The small pores, or the bulk volume of small pores, which hold water in soils against a tension usually >60 cm of water. See also **water tension**.

capillary potential–(no longer used in SSSA publications) As originally proposed by E. Buckingham in 1907, the definition was unconventional with respect to sign, being the negative of the matric potential. See also **Table 5. Soil water terms**.

capillary water–(no longer used in SSSA publications) The water held in the "capillary" or small pores of a soil, usually with a tension >60 cm of water. See also **water tension**.

carbamoylurea–See **biuret**.

carbon cycle–The sequence of transformations whereby carbon dioxide is converted to organic forms by photosynthesis or chemosynthesis, recycled through the biosphere (with partial incorporation into sediments), and ultimately returned to its original state through respiration or combustion.

carbon-nitrogen ratio–See **carbon-organic nitrogen ratio**.

carbon-organic nitrogen ratio–The ratio of the mass of organic carbon to the mass of organic nitrogen in soil, organic material, plants, or microbial cells.

Carolina Bay–Any of various shallow, often oval or elliptical, generally marshy, closed depressions in the Atlantic coastal plain (from southern New Jersey to northeastern Florida, especially developed in the Carolinas). They range from about 100 meters to many kilometers in length, are rich in humus, and under native conditions contain trees and shrubs different from those of the surrounding areas.

cartographic unit–See **map unit, soil**; **soil map**.

cat clay–Poorly drained, clayey soils, commonly formed in an estuarine environment, that become very acidic when drained due to oxidation of ferrous sulfide.

catabolism–The breakdown of organic compounds within an organism.

catch crop–(i) A crop produced incidental to the main crop of the farm and usually occupying the land for a short period. (ii) A crop grown to replace a main crop that has failed.

category–Any one of the ranks of the system of soil classification in which soils are grouped on the basis of their characteristics.

catena–(as used in the USA) A sequence of soils of about the same age, derived from similar parent material, and occurring under similar climatic conditions, but having different characteristics due to variation in relief and in drainage. See also **toposequence**.

cation exchange–The interchange between a cation in solution and another cation in the boundary layer between the solution and surface of negatively charged material such as clay or organic matter.

cation exchange capacity (CEC)– The sum of **exchangeable bases** plus total soil acidity at a specific pH, values, usually 7.0 or 8.0. When acidity is expressed as **salt extractable acidity**, the cation exchange capacity is called the effective cation exchange capacity (ECEC) because this is considered to be the CEC of the exchanger at the native pH value. It is usually expressed in centimoles of charge per kilogram of exchanger (cmol$_c$ kg^{-1}) or millimoles of charge per kilogram of exchanger. See also **acidity, total**.

cavitation–The formation of gas or water vapor-filled cavities in a liquid volume when the pressure is reduced (tension is increased) to a critical level. In water systems cavitation typically occurs at about 0.08 MPa of water tension. In confined systems, cavitation can create discontinuity of water columns preventing the non-elastic transmission of pressure along the column across the cavitation.

cemented–Having a hard, brittle consistency because the particles are held together by cementing substances such as humus, $CaCO_3$, or the oxides of silicon, iron, and aluminum. The hardness and brittleness persist even when wet. See also **consistence**.

center-pivot–See **irrigation**, *center-pivot*.

chambers–**Vesicles** or **vughs** connected by a **channel** or **channels**.

channel–(i) A tubular-shaped void. (ii) A natural stream that conveys water; a ditch excavated for the flow of water.

channer–In Scotland and Ireland, gravel; in the USA, thin, flat rock fragments up to 150 mm on the long axis. See also **rock fragments**.

channery–See **rock fragments**.

check-basin–See **irrigation**, *check-basin*.

chelates–Organic chemicals with two or more functional groups that can bind with metals to form a ring structure. Soil organic matter can form chelate structures with some metals, especially transition metals, but, much metal ion binding in soil organic matter probably does not involve chelation. Artificial chelating compounds are sometimes added to soil to increases the soluble fraction of some metals.

chemical fallow–See **tillage**, *chemical fallow*.

chemically precipitated phosphorus–(no longer used in SSSA publications) Relatively insoluble phosphorus compounds resulting from reactions of phosphorus with soil constituents: e. g. calcium and magnesium phosphates which are precipitated above a pH of about 6.0 to 6.5 (if calcium and magnesium are present); and, iron and aluminum phosphates which are precipitated below a pH of about 5.8 to 6.1. See also **phosphorus fixation**.

chemical oxygen demand (COD)– A measure of the oxygen-consuming capacity of inorganic and organic matter present in water or wastewater. The COD test, like the **BOD** test, is used to determine the degree of pollution in an effluent.

chemical potential–(i) The rate of change of Gibbs free energy, G, with respect to the number of moles of one component in a mixed chemical system at fixed temperature, pressure and number of moles of other components. (ii) The chemical potential of a component increases with increasing concentration or partial pressure. See also **activity (chemical)**.

chemical weathering–The breakdown of rocks and minerals due to the presence of water and other components in the **soil solution** or changes in **redox potential**. See also **weathering**.

chemigation–The process where **fertilizers**, **pesticides** and other agrichemicals are applied into irrigation water to fertilize crops, control pests or amend soils.

chemisorbed phosphorus–(no longer used in SSSA publications) Phosphorus adsorbed or precipitated on the surface of clay minerals or other crystalline materials. See also **adsorption**, **chemically precipitated phosphorus**, and **phosphorus fixation**.

chemodenitrification–Nonbiological processes leading to the production of gaseous forms of nitrogen (molecular nitrogen or an oxide of nitrogen).

chemolithotroph–An organism capable of using CO_2 or carbonates as the sole source of carbon for cell biosynthesis, and deriving energy from the oxidation of reduced inorganic or organic compounds. Used synonymously with "chemolithoautotroph" and "chemotroph."

chemoorganotroph–An organism for which organic compounds serve as both energy and carbon sources for cell synthesis. Used synonymously with "**heterotroph**."

chemostat–A device for the continuous culture of microorganisms in which growth rate and population size are regulated by the concentration of a limiting nutrient in incoming medium.

chemotaxis–The oriented movement of a motile organism with reference to a chemical agent. May be positive (toward) or negative (away) with respect to the chemical gradient.

Chernozem–A **zonal** great soil group consisting of soils with a thick, nearly black or black, organic matter-rich A horizon high in exchangeable calcium, underlain by a lighter colored transitional horizon above a zone of calcium carbonate accumulation; occurs in a cool subhumid climate under a vegetation of tail and midgrass prairie. (Not used in current U.S. system of soil taxonomy.)

Chestnut soil–A **zonal** great soil group consisting of soils with a moderately thick, dark-brown A horizon over a lighter colored horizon that is above a zone of calcium carbonate accumulation. (Not used in current U.S. system of soil taxonomy.)

chisel–See **tillage**, *chisel*.

chlorite–A group of layer silicate minerals of the 2:1 type that has the interlayer filled with a positively charged metal-hydroxide octahedral sheet. There are both trioctahedral (e.g., M = Fe(II), Mg^{2+}, Mn^{2+}, Ni^{2+}) and dioctahedral (M= Al^{3+}, Fe^{3+}, Cr^{3+}) varieties. See also **Appendix I, Table A3**.

chopping–A method of preparing forest soils for planting or seeding by passing a heavy drum roller with sharp parallel blades over the site to break up organic debris and mix it into the mineral soil.

chroma–The relative purity, strength, or saturation of a color; directly related to the dominance of the determining wavelength of the light and inversely related to grayness; one of the three variables of color. See also **Munsell color system**, **hue**, and **value**.

chronosequence–A group of related soils that differ, one from the other, primarily as a result of differences in time as a soil-forming factor.

cinder land–In Soil Survey a map unit that is a type of **miscellaneous area**, which is composed of loose cinders and other **pyroclastic** materials.

cinders–Uncemented vitric, vesicular, **pyroclastic** material, more than 2.0 mm in at least one dimension, with an apparent specific gravity (including vesicles) of more than 1.0 and less than 2.0.

cirque–Semicircular, concave, bowl-like area with steep face primarily resulting from erosive activity of a mountain glacier.

cirque land–In Soil Survey, a map unit that is a type of **miscellaneous area**, which consists of areas of rock and rubble in a **cirque** basin.

citrate-soluble phosphorus–The fraction of total P in fertilizer that is insoluble in water but soluble in neutral 0.33 M ammonium citrate. Together with **water-soluble phosphate**, this represents the readily available P content of the fertilizer. See also **phosphate**.

class, soil–A group of soils defined as having a specific range in one or more particular property(ies) such as acidity, degree of slope, texture, structure, land-use capability, degree of erosion, or drainage. See also **soil texture** and **soil structure**.

classification, soil–The systematic arrangement of soils into groups or categories on the basis of their characteristics. Broad groupings are made on the basis of general characteristics and subdivisions on the basis of more detailed differences in specific properties. The USDA soil classification system of soil taxonomy was adapted for use in publications by the National Cooperative Soil Survey on 1 Jan. 1965. Abridged statements of diagnostic features, orders, and suborders are listed alphabetically. The outline of the system is shown in **Appendix I (Table A1)**. Great groups are named by adding a prefix to the suborder name. A list of the connotations of these prefixes is shown in **Appendix I (Table A2)**. For complete definitions of taxa see: Soil Survey Staff, 1994, *Keys to soil taxonomy*, 6th Edition, U.S. Government Printing Office.

clastic–Pertaining to rock or sediment composed mainly of fragments derived from preexisting rocks or minerals and moved from their place of origin. The term indicates sediment sources that are both within and outside the depositional basin.

clay–(i) A soil separate consisting of particles <0.002 mm in equivalent diameter. See also **soil separates**. (ii) A textural class. See also **soil texture**. (iii) (In reference to clay mineralogy) A naturally occurring material composed primarily of fine-grained minerals, which is generally plastic at appropriate water contents and will harden when dried or fired. Although clay usually contains phyllosilicates, it may contain other materials that impart plasticity and harden when dried or fired. Associated phases in clay may include materials that do not impart plasticity and organic matter.

clay coating–Same as **clay film**.

clay films–Coatings of oriented clay on the surfaces of peds and mineral grains and lining pores. Also called **clay skins**, **clay flows**, **illuviation cutans**, or **argillans**.

clay flows–See **clay films**.

clay loam–A soil textural class. See also **soil texture**.

clay mineral–A phyllosilicate mineral or a mineral that imparts plasticity to **clay** and which harden upon drying or firing. See also **phyllosilicate mineral terminology**.

clay mineralogy–See **phyllosilicate mineral terminology**.

clay skins–See **clay films**.

clayey–(i) Texture group consisting of sandy clay, silty clay, and clay soil textures. See also **soil texture**. (ii) Family particle-size class for soils with 35% or more clay and <35% rock fragments in upper subsoil horizons.

claypan–A dense, compact, slowly permeable layer in the subsoil having a much higher clay content than the overlying material, from which it is separated by a sharply defined boundary. Claypans are usually hard when dry, and plastic and sticky when wet.

climatic index–A simple, single numerical value that expresses climatic relationships; for example, the numerical value obtained in Transeau's precipitation-evaporation ratio.

climax–(no longer used in SSSA publications) The most advanced successional community of plants capable of development under, and in dynamic equilibrium with, the prevailing environment.

climosequence–A group of related soils that differ, one from another, primarily as a result of differences in climate as a soil-forming factor.

clod–A compact, coherent mass of soil varying in size, usually produced by plowing, digging, etc., especially when these operations are performed on soils that are either too wet or too dry and usually formed by compression, or breaking off from a larger unit, as opposed to a building-up action as in **aggregation**.

coarse fragments–See **rock fragments**.

coarse sand–(i) A soil separate. See also **soil separates**. (ii) A soil textural class. See also **soil texture**.

coarse sandy loam–A soil textural class. See also **soil texture**.

coarse textured–Texture group consisting of sand and loamy sand textures. See also **soil texture**.

coastal plain–Any plain of unconsolidated fluvial or marine sediment which had its margin on the shore of a large body of water, particularly the sea, e.g., the coastal plain of the Southeastern USA, extending for 5000 km from New Jersey to Texas.

coating–A layer of a substance completely or partly covering a surface of soil material. Coatings include clay coatings, calcite coatings, gypsum coatings, organic coatings, salt coatings, etc.

cobbles–See **cobblestones**

cobblestones–Rounded or partially rounded rock or mineral fragments between 75 and 250 mm in diameter. See also **rock fragments**

cobbly–Containing appreciable quantities of **cobblestones**. See also **rock fragments**.

COD–See **chemical oxygen demand**.

coefficient of linear extensibility (COLE)–The percent shrinkage in one dimension of a molded soil between two water contents, e.g. between its plastic limit to air dry.

coliform–A general term for a group of bacteria that inhabit the intestinal tract of humans and other animals. Their presence in water constitutes presumptive evidence for fecal contamination. Includes all aerobic and facultatively anaerobic, gram-negative rods that are nonspore forming and that ferment lactose with gas formation. *Escherichia coli* and *Enterobacter* are important members.

colloid–A particle, which may be a molecular aggregate, with a diameter of 0.1 to 0.001 μm. Soil clays and soil organic matter are often called soil colloids because they have particle sizes that are within, or approach colloidal dimensions.

colloidal suspension–Suspension in water of particles so finely divided that they will not settle under the action of gravity, but will diffuse, even in quiet water, under the random impulses of Brownian motion. Particle sizes range from about 1 μm to about 1nm; however, there is no sharp differentiation by size between coarse ("true") suspension and colloidal suspension or between colloidal suspension and solution.

colluvial–Pertaining to material or processes associated with transportation and/or deposition by mass movement (direct gravitational action) and local, unconcentrated runoff on side slopes and/or at the base of slopes.

colluvium–Unconsolidated, unsorted earth material being transported or deposited on sideslopes and/or at the base of slopes by mass movement (e.g., direct gravitational action) and by local, unconcentrated runoff.

colonization–Establishment of a community of microorganisms at a specific site or ecosystem.

color–See **Munsell color system**.

color composite (multiband photography)–A color picture produced by assigning a color to a particular spectral band. Ordinarily blue is assigned to band 1 or 4 (~ 500 to 600 nm), green to band 2 or 5 (~600 to 700 nm), and red to band 3 (~700 nm to 1 μm) or 7 (~800 nm to 1.1 μm), to form a picture closely approximating a color-infrared photograph.

colter slit–See **tillage**.

columnar soil structure–A shape of soil structure. See also **soil structure** and **soil structure shapes**.

cometabolism–Transformation of a substrate by a microorganism without deriving energy, carbon, or nutrients from the substrate. The microorganism can transform the substrate into intermediate degradation products but fails to multiply.

commensalism–Interaction between two species in which one species derives benefit while the other is unaffected.

community–All of the organisms that occupy a common habitat and that interact with one another.

compaction–(i) To unite firmly; the act or process of becoming compact. (ii) (geology) The changing of loose sediment into hard, firm rock. (iii) (soil engineering) The process by which the soil grains are rearranged to decrease void space and bring them into closer contact with one another, thereby increasing the **bulk density**. (iv) (solid waste disposal) The reducing of the bulk of solid waste by rolling and tamping.

competence–The ability of a current of water or wind to transport sediment, in terms of particle size rather than amount, measured as the diameter of the largest particle transported. It depends upon velocity: a small but swift current for example, may have greater competence than a larger but slower moving current.

competition–A rivalry between two or more species for a limiting factor in the environment.

complex, soil–See **soil complex**.

component soil–The collection of **polypedons** or parts of **polypedons** within a map unit that are members of the **taxon** (or a kind of miscellaneous unit) for which the map unit is named. Simple or complex names for the component soils are formed from a class name (taxon name) from some categorical level of the U.S. system of soil taxonomy, with or without an additional phase identification for utilitarian features. See also **inclusion** and **map unit**.

compost–Organic residues, or a mixture of organic residues and soil, that have been mixed, piled, and moistened, with or without addition of fertilizer and lime, and generally allowed to undergo **thermophilic** decomposition until the original organic materials have been substantially altered or decomposed. Sometimes called "artificial manure" or "synthetic manure." In Europe, the term may refer to a potting mix for container-grown plants.

composting–A controlled biological process which converts organic constituents, usually wastes, into humus-like material suitable for use as a soil amendment or organic fertilizer.

compressibility–The property of a soil pertaining to its susceptibility to decrease in bulk volume when subjected to a load.

compressibility index–The ratio of pressure to void ratio on the linear portion of the curve relating the two variables.

concentrated flow–A relatively large water flow over or through a relatively narrow course.

concentration–The amount of suspended or dissolved particles, or elements in a unit volume or unit mass as specified at a given temperature and pressure.

concretion–(i) A cemented concentration of a chemical compound, such as calcium carbonate or iron oxide, that can be removed from the soil intact and that has crude internal symmetry organized around a point, line, or plane. (ii) (micromorphological) A **glaebule** with a generally concentric fabric about a center which may be a point, line, or a plane.

conductivity, hydraulic–See **soil water**, *hydraulic conductivity*.

cone index–The force per unit basal area required to push a cone penetrometer through a specified increment of soil. See also **cone penetrometer**.

cone penetrometer–An instrument in the form of a cylindrical rod with a cone-shaped tip designed for penetrating soil and for measuring the end-bearing component of **penetration resistance**. The resistance to penetration developed by the cone equals the vertical force applied to the cone divided by its horizontally projected area. See also **cone index, penetration resistance** and **friction cone penetrometer**.

conformity–The mutual and undisturbed relationship between adjacent sedimentary strata that have been deposited in orderly sequence with little or no evidence of time lapses; true stratigraphic continuity in the sequence of beds without evidence that the lower beds were folded, tilted, or eroded before the higher beds were deposited.

conjugated metabolites–Metabolically produced compounds that are linked together by covalent binding (complex formation).

conjunctive water use–See **irrigation**, *conjunctive water use*.

consistence–The attributes of soil material as expressed in degree of cohesion and adhesion or in resistance to deformation or rupture. See **Table 1**.

consistency–The manifestations of the forces of cohesion and adhesion acting within the soil at various water contents, as expressed by the relative ease with which a soil can be deformed or ruptured. Engineering descriptions include: (i) the designation of five inplace categories (soft, firm or medium, stiff, very stiff, and hard) as assessed by thumb and thumbnail penetrability and indentability; and (ii) characterization by the Atterberg limits (i.e., liquid limit, plastic limit, and plasticity number). See also **Atterberg limits, liquid limit, plastic limit**, and **plasticity number**.

Table 1. Terms for describing consistence (rupture resistance) of blocklike specimens. (From: Soil survey division staff. 1993. *Soil survey manual*, USDA-SCS Agric. Handb. 18. p. 174-175. U.S. Gov. Print. Office, Washington, DC.

Classes for moisture states			Test description	
Moderately dry and very dry	Slightly dry and wetter	Air dry, submerged	Operation	Stress applied
Loose	Loose	Not applicable	Specimen not obtainable	
Soft	Very friable	Non-cemented	Fails under very slight force applied slowly between thumb and forefinger	<8 N
Slightly hard	Friable	Extremely weakly cemented	Fails under slight force applied slowly between thumb and forefinger	8-20 N
Moderately hard	Firm	Very weakly cemented	Fails under moderate force applied slowly between thumb and forefinger	20-40 N
Hard	Very firm	Weakly cemented	Fails under strong force applied slowly between thumb and forefinger	40-80 N
Very hard	Extremely firm	Moderately cemented	Cannot be failed between thumb and forefinger but can be between both hands or by placing on a nonresilent surface and applying gentle force underfoot	80-160 N
Extremely hard	Slightly rigid	Strongly cemented	Cannot be failed in hands but can be underfoot by full body weight applied slowly	160-800 N
Rigid	Rigid	Very strongly cemented	Cannot be failed underfoot by full body weight but can be by <300 J blow	800N-300 J
Very rigid	Very rigid	Indurated	Cannot be failed by blow of <300 J	$\geq$300 J

consolidation test–A test in which the soil specimen is laterally confined in a ring and is compressed between porous plates.

constant-charge surface–A mineral surface carrying a net electrical charge whose magnitude depends only on the structure and chemical composition of the mineral itself. Constant charge surfaces in soils usually arise from **isomorphous substitution** in **phyllosilicate** clay structures.

constant-potential surface–Variable charge surfaces are called constant potential surfaces because at constant activity of the potential determining ion (e.g. constant pH) the electrical potential difference between the solid surface and the bulk solution is constant. See also **variable charge, pH-dependent charge**, and **constant-charge surface**.

constructional surface–A land surface owing its origin and form to depositional processes, with little or no modification by erosion.

consumptive irrigation requirement–See **irrigation**, *consumptive irrigation requirement*.

contact angle–Where water is in contact with a solid surface, the angle occurring on the liquid side of a meniscus or water droplet between the flat solid surface and the gas phase beyond the liquid.

continuous delivery–See **irrigation**, *continuous delivery*.

continuous permafrost–Permafrost occurring everywhere beneath the exposed land surface throughout a geographic region. See also **permafrost**.

contour ditch–See **irrigation**, *contour ditch*.

contour flooding–See **irrigation**, *contour flooding*.

contour strip cropping–See **tillage**, *strip cropping*.

contrasting soil–A soil that does not share diagnostic criteria and does not behave or perform similar to the soil being compared.

controlled drainage–See **irrigation**, *controlled drainage*.

controlled traffic–See **tillage**, *controlled traffic*.

convection–A process by which heat, solutes, or particles are transferred from one part of a fluid to another by movement of the fluid itself; also called **advection**.

conveyance loss–See **irrigation**, *conveyance loss*.

copiotrophs–See **zymogenous flora.**

coppice mound–See **shrub-coppice dune**.

coprogenic material–Remains of fish excreta and similar materials that occur in some organic soils.

corrugate–See **irrigation**, *corrugate*.

cover crop–Close-growing crop, that provides soil protection, seeding protection, and soil improvement between periods of normal crop production, or between trees in orchards and vines in vineyards. When plowed under and incorporated into the soil, cover crops may be referred to as **green manure crops**.

cradle knoll–See **tree-tip pit** and **tree-tip mound**.

creep–Slow **mass movement** of soil and soil material down slopes driven primarily by gravity, but facilitated by saturation with water and by alternate freezing and thawing.

crest–See **summit**.

critical nutrient concentration–The nutrient concentration in the plant, or specified plant part, above which additional plant growth response slows. Crop yield, quality or performance are less than optimum when the concentration is less.

critical soil test concentration–That concentration at which 95% of maximum **relative yield** is achieved. Usually coincides with the inflection point of a curvilinear yield response curve.

crop nutrient requirement–The amount of nutrients needed to grow a specified yield of a crop plant per unit area.

crop residue management–See **tillage,** *crop residue management.*

crop residue management system–See **tillage,** *crop residue management system.*

crop rotation–A planned sequence of crops growing in a regularly recurring succession on the same area of land, as contrasted to continuous culture of one crop or growing a variable sequence of crops.

cross cultivation–See **tillage,** *cross cultivation.*

cross-slope bench–See **terrace.**

cross-stratification–Arrangement of strata inclined at an angle to the main stratification. This is a general term having two subdivisions; cross-bedding, in which the cross-strata are thicker than 1 cm, and cross-lamination, in which they are thinner than 1 cm. A single group of related cross-strata is a set and a group of similar, related sets is a coset.

crotovina–(no longer used in SSSA publications) A former animal burrow in one soil horizon that has been filled with organic matter or material from another horizon (also spelled "krotovina").

crumb (aggregate)–A soft, porous, more or less rounded ped from 1 to 5 mm in diameter. (Not used in current U.S. system of soil taxonomy.) See also **soil structure shapes** and **Table 1.**

crumb structure–A structural condition in which most of the peds are crumbs. (Not used in current U.S. system of soil taxonomy.) See also **soil structure shapes.**

crushing–See **tillage,** *crushing.*

crushing strength–The force required to crush a mass of dry soil or, conversely, the resistance of the dry soil mass to crushing. Expressed in units of force per unit area (pressure).

crust–A transient soil-surface layer, ranging in thickness from a few millimeters to a few centimeters, that is either denser, structurally different or more cemented than the material immediately beneath it, resulting in greater **soil strength** when dry as measured by **penetration resistance** or other indices of soil strength.

Cryands–Andisols that have a **cryic** or **pergelic** soil temperature regime. (A suborder in the U.S. system of soil taxonomy.)

Cryerts–Vertisols that have a **cryic** soil temperature regime. (A suborder in the U.S. system of soil taxonomy.)

cryic–A soil temperature regime that has mean annual soil temperatures of $>0°$ C but $<8°$ C, $>5°$ C difference between mean summer and mean winter soil temperatures at 50 cm, and cold summer temperatures.

Cryids–Aridisols that have a **cryic** soil temperature regime. (A suborder in the U.S. system of soil taxonomy.)

Cryods–Spodosols that have a **cryic** or **pergelic** soil temperature regime. (A suborder in the U.S. system of soil taxonomy.)

calcareous material; formed in arid regions under sparse shrub vegetation. (Not used in current U.S. system of soil taxonomy.)

desert varnish–A thin, dark, shiny film or coating of iron oxide and lesser amounts of manganese oxide and silica formed on the surfaces of pebbles, boulders, rock fragments, and rock outcrops in arid regions.

desorption–The migration of adsorbed entities off of the adsorption sites. The inverse of **adsorption**.

detachment–See **erosion**, *detachment.*

detoxification–Conversion of a toxic molecule or ion into a nontoxic form.

detritus–Dissolved and particulate dead **organic matter**. See also **coprogenic material**.

diagnostic horizons–(as used in the U.S. system of soil taxonomy) Combinations of specific soil characteristics that are indicative of certain classes of soils. Those which occur at the soil surface are called epipedons, those below the surface, diagnostic subsurface horizons.

diatomaceous earth–A geologic deposit of fine, grayish siliceous material composed chiefly or wholly of the remains of **diatoms**. It may occur as a powder or as a porous, rigid material.

diatoms–Algae having siliceous cell walls that persist as a skeleton after death. Any of the microscopic unicellular or colonial algae constituting the class *Bacillariaceae*. They are abundant in fresh and salt waters and their remains are widely distributed in soils.

diazotroph–A microorganism or association of microorganisms that can reduce molecular nitrogen (N_2) to ammonia.

differential water capacity–See **soil water**, *differential water capacity.*

diffuse double layer–A conceptual model of a heterogeneous system that consists of a solid surface (e.g. clay or oxide surface) having a net electrical charge together with an ionic swarm in solution containing ions of opposite charge, neutralizing the surface charge.

diffusion (nutrient)–The movement of nutrients in soil because of a **chemical activity** gradient.

dig–See **tillage**, *dig.*

digestibility–(as applied to organic wastes) The potential degree to which organic matter in waste water or sewage can be broken down into simpler and/or more biologically stable products.

dinitrogen fixation–Conversion of molecular nitrogen (N_2) to ammonia and subsequently to organic nitrogen utilizable in biological processes.

dioctahedral–An octahedral sheet or a mineral containing such a sheet that has two-thirds of the octahedral sites filled by trivalent ions such as aluminum or ferric iron. See also **phyllosilicate mineral terminology**, **trioctahedral** and **Appendix I, Table A3**.

dip–The angle that a structural surface, e.g. a bedding or fault plane, makes with the horizontal, measured perpendicular to the strike of the structure and in the vertical plane.

direct counts–In soil microbiology, a method of estimating the total number of microorganisms in a given mass of soil by direct microscopic examination.

direct problem–The predicting of the behavior of a system given its inherent properties.

direct shear test–A shear test in which soil under an applied normal load is stressed to failure by moving one section of the sample or sample container relative to the other section.

discharge curve–See **irrigation**, *discharge curve*.

discontinuity–Any interruption in sedimentation, whatever its cause or length, usually a manifestation of nondeposition and accompanying erosion.

discontinuous permafrost–**Permafrost** occurring in some areas beneath the exposed land surface throughout a geographic region where other areas are free of permafrost. See also **continuous permafrost, sporadic permafrost**.

disintegration–See **mechanical weathering**.

dispersion–(i) A term used in relation to solute movement. See also **hydrodynamic dispersion**. (ii) The break-down of soil aggregates into individual component particles. See also **deflocculation**.

dispersivity–The ratio of the hydrodynamic dispersion coefficient (d) divided by the pore water velocity (v); thus D= d/v.

dissection–Fluvial erosion of a land surface or landform by the cutting of gullies, arroyos, canyons, or valleys leaving ridges, hills, mountains or flat-topped remnants separated by drainageways.

dissimilation–The release from cells of inorganic or organic substances formed by metabolism.

distal–Said of a sedimentary deposit consisting of fine **clastics** and deposited farthest from the source area.

distribution coefficient (K_d)–The distribution of a chemical between soil and water.

diversion dam–A structure or barrier built to divert part or all of the water of a stream to a different course.

divide–The line of separation, or the summit area, or narrow tract of higher ground that constitutes the watershed boundary between two adjacent drainage basins; it divides the surface waters that flow naturally in one direction from those that flow in the opposite direction.

dolomitic lime–A naturally occurring liming material composed chiefly of carbonates of Mg and Ca in approximately equimolar proportions.

drag–(i) The force retarding the flow of a fluid over the surface of a solid body. (ii) See **tillage**, *drag*.

drain tile–Concrete, ceramic, plastic, or other rigid pipe or similar buried structure used to collect and conduct profile drain-water from the soil in a field.

drain, to–(i) To provide channels, such as open ditches or drain tile, so that excess water can be removed by surface or by internal flow. (ii) To lose water (from the soil) by percolation.

drainage basin–A general term for a region or area bounded by a drainage divide and occupied by a drainage system.

drainage class (natural)–A group of soils defined as having a specific range in relative wetness under natural conditions as it pertains to wetness due to a water table under conditions similar to those under which the soil developed.

drainage curves–See **irrigation**, *drainage curves*.

drainage pattern–The configuration of arrangement in plan view of the natural stream courses in an area. It is related to local geologic and geomorphologic features and history.

drainage terrace–See **terrace**.

drainage, excessive–Too much or too rapid loss of water from soils, either by percolation or by surface flow. The occurrence of internal free water is very rare or very deep.

drainage, surface–Used to refer to surface movement of excess water–includes such terms as ponded, flooded, slow, and rapid.

drainageway–A general term for a course or channel along which water moves in draining an area.

drift–See **glacial drift**.

drip irrigation–See **irrigation**.

drumlin–A low, smooth, elongated oval hill, mound, or ridge of compact till that may or may not have a core of bedrock or **stratified** drift. The longer axis is parallel to the general direction of **glacier** flow. Drumlins are products of streamline (laminar) flow of glaciers, which molded the subglacial floor through a combination of erosion and deposition.

dryland farming–Crop production without irrigation (rainfed agriculture).

dry-mass content or ratio–The ratio of the mass of any component (of a soil) to the oven-dry mass of the soil. See also **oven-dry soil**.

dry-weight percentage–See **dry-mass content or ratio**.

duff–See **litter**.

duff mull–A forest humus type, transitional between mull and mor, characterized by an accumulation or organic matter on the soil surface in friable Oe horizons, reflecting the dominant zoogenous decomposers. They are similar to **mors** in that they generally feature an accumulation of partially to well-humified organic materials resting on the mineral soil. They are similar to **mulls** in that they are zoologically active. **Duff mulls** usually have four horizons: Oi(L), Oe(F), Oa(H), and A. Sometimes differentiated into the following Groups: Mormoder, Leptomoder, Mullmoder, Lignomoder, Hydromoder, and Saprimoder.

dumps–Areas of smooth or uneven accumulations or piles of waste rock or general refuse that without major reclamation are incapable of supporting plants.

dune–A low mound, ridge, bank or hill of loose, windblown, granular material (generally sand), either bare or covered with vegetation, capable of movement from place to place but always retaining its characteristic shape.

dune land–In Soil Survey a map unit that is a type of miscellaneous area, which consists of sand dunes and intervening troughs that shift with the wind. See also **miscellaneous area**.

Durids–**Aridisols** which have a **duripan** that has its upper boundary within 100 cm of the soil surface. (A suborder in the U.S. system of soil taxonomy.)

durinodes–Weakly cemented to indurated soil nodules cemented with SiO_2. Durinodes break down in concentrated KOH after treatment with HCl to remove carbonates, but do not break down on treatment with concentrated HCl alone.

duripan–A subsurface soil horizon that is cemented by illuvial silica, usually opal or microcrystalline forms of silica, to the degree that less than 50 percent of the volume of air-dry fragments will slake in water or HCl.

dust mulch–A very loose, finely granular, or powdery condition on the soil surface.

dy–Colloidal humic substances that accumulate in peat soils at the transition zone between the peat and the underlying mineral material.

dynamic head–See **irrigation**, *dynamic head.*

dynamic penetrometer–A penetrometer which is driven into the soil by a hammer or falling weight.

dysic–Low level of bases in soil material, specified at family level of classification.

E

E horizon–See **soil horizon** and **Appendix II**.

EC$_e$–The electrical conductance of an extract from a soil saturated with distilled water, normally expressed in units of siemens or decisiemens per meter at 25° C.

ecofallow–See **tillage**, *chemical fallow.*

economic rate–The application rate of material, usually **fertilizer**, that gives the highest economic returns for the crop produced.

EC–See **electrical conductivity**.

ectomycorrhiza(e)–A **mycorrhizal** association in which the fungal **mycelia** extend inward, between root cortical cells, to form a network ("Hartig net") and outward into the surrounding soil. Usually the fungal hyphae also form a mantle on the surface of the roots.

edaphic–(i) Of or pertaining to the soil. (ii) Resulting from or influenced by factors inherent in the soil or other substrate, rather than by climatic factors.

edaphology–The science that deals with the influence of soils on living things; particularly plants, including man's use of land for plant growth.

effective cation exchange capacity (ECEC)–See **cation exchange capacity(CEC)**.

effective precipitation–That portion of the total rainfall precipitation which becomes available for plant growth.

effective stress–The stress transmitted through a soil by intergranular pressures.

E$_H$–The potential that is generated between an oxidation or reduction half-reaction and the standard hydrogen electrode (0.0v at pH = 0). In soils it is the potential created by oxidation-reduction reactions that take place on the surface of a platinum electrode measured against a reference electrode minus the E$_H$ of the reference electrode . This is a measure of the oxidation-reduction potential of electrode reactive components in the soil. See also **pe**.

electrical conductivity (EC)–Conductivity of electricity through water or an extract of soil. Commonly used to estimate the soluble salt content in solution.

electrokinetic (zeta) potential–The electrical potential at the surface of the shear plane between immobile liquid attached to a charged particle and mobile liquid further from the particle surface.

electron acceptor–A compound which accepts electrons during biotic or abiotic chemical reactions and is thereby reduced.

electron donor–A compound which donates or supplies electrons during metabolism and is thereby oxidized.

eluvial horizon–A soil horizon that has been formed by the process of eluviation. See also **illuvial horizon**.

eluviation–The removal of soil material in suspension (or in solution) from a layer or layers of a soil. Usually, the loss of material in solution is described by the term "leaching." See also **illuviation** and **leaching**.

emitter–See **irrigation**, *trickle irrigation, emitter*.

end moraine–A ridge-like accumulation that is being or was produced at the outer margin of an actively flowing **glacier** at any given time; a **moraine** that has been deposited at the outer or lower end of a valley glacier.

endomycorrhiza–A **mycorrhizal** association with intracellular penetration of the host root cortical cells by the fungus as well as outward extension into the surrounding soil. See also **arbuscule**, **vesicles**, and **vesicular arbuscular**.

endophyte–An organism (e.g., fungus, bacteria) growing within a plant. The association may be symbiotic or parasitic.

endosaturation–The soil is saturated with water in all layers from the upper boundary of saturation to a depth of 200 cm or more from the mineral soil surface. See also **episaturation**.

enrichment culture–A technique in which environmental (including nutritional) conditions are controlled to favor the development of a specific organism or group of organisms through prolonged or repeated culture.

enrichment ratio (ER)–See **erosion**, *enrichment ratio (ER)*.

Entisols–Mineral soils that have no distinct subsurface diagnostic horizons within 1 m of the soil surface. (An order in the U.S. system of soil taxonomy.)

enzyme–Any of numerous proteins that are produced in the cells of living organisms and function as catalysts in the chemical processes of those organisms.

eolian–Pertaining to earth material transported and deposited by the wind including **dune** sands, **sand sheets**, **loess**, and **parna**.

ephemeral gully–See **erosion**, *ephemeral gully*.

ephemeral stream–A stream, or reach of a stream, that flows only in direct response to precipitation. It receives no protracted supply from melting snow or other source, and its channel is, at all times, above the water table.

epipedon–See **diagnostic horizon**.

episaturation–The soil is saturated with water in one or more layers within 200 cm of the mineral soil surface and also has one or more unsaturated layers with an upper boundary above 200 cm depth, below the saturated layer(s) (a perched water table). See also **endosaturation**.

episomes–See **plasmids**.

equation of continuity–An equation expressing the conservation of mass or energy as it applies to soil water, heat, air, etc., moving through soil.

equivalent diameter–In sedimentation analysis of particle size, the diameter assigned to a nonspherical particle; that is, the diameter of a spherical particle of the same density and velocity of fall. Sometimes referred to as the **equivalent spherical diameter**.

equivalent spherical diameter–See **equivalent diameter**

erodibility–See **erosion**, *erodibility.*

erodible–See **erosion**, *erodible.*

erosion–(i) The wearing away of the land surface by rain or irrigation water, wind, ice, or other natural or anthropogenic agents that abrade, detach and remove geologic parent material or soil from one point on the earth's surface and deposit it elsewhere, including such processes as gravitational **creep** and so-called tillage erosion; (ii) The detachment and movement of soil or rock by water, wind, ice, or gravity. The following terms are used to describe different erosion types, processes, and mechanisms.

accelerated erosion–Erosion in excess of natural rates, usually as a result of anthropogenic activities.

bed load–The sediment that moves by sliding, rolling, or salting on or very near the streambed; sediment moved mainly by tractive or gravitational forces or both but at velocities less than the surrounding flow.

detachment–The removal of transportable fragments of soil material from a soil mass by an eroding agent, usually falling raindrops, running water, or wind; through detachment, soil particles or aggregates are made ready for transport.

enrichment ratio (ER)–The ratio of a compound's concentration in the eroded soil to the noneroded soil; the same for eroded water flow to the normal water flow.

ephemeral gully–Small channels eroded by concentrated flow that can be easily filled by normal tillage, only to reform again in the same location by additional runoff events.

erodibility–(i)The degree or intensity of a soil's state or condition of, or susceptibility to, being erodible. (ii) The K factor in the **Universal Soil Loss Equation**. See also **erosion**.

erodible–Susceptible to erosion.

erosion classes–A grouping of erosion conditions based on the degree of erosion or on characteristic patterns. (Applied to accelerated erosion; not to normal, natural, or geological erosion.) Four erosion classes are recognized for water erosion and three for wind erosion. Specific definitions for each vary somewhat from one climatic zone, or major soil group, to another. (For details see: Soil survey division staff. 1993. *Soil survey manual*, USDA-SCS Agric. Handb. 18. U.S. Gov. Print. Office, Washington, DC.)

erosion pavement–A layer of coarse fragments, such as sand or gravel, remaining on the surface of the ground after the removal of fine particles by erosion. See also **desert pavement**.

erosion potential (EI) –A numerical value expressing the inherent erodibility of a soil or maximum potential erosion. In the Universal Soil Loss Equation (under clean tillage, up and down slope) EI = *RKLS/T.*

erosional surface–A land surface shaped by the erosive action of ice, wind, or water; but usually as the result of running water.

erosive velocity–Velocity of the erosive agent necessary to cause erosion.

erosivity–The measured or predicted ability of water, wind, gravity, or any other erosion agent, to cause erosion.

essential [chemical] elements–Elements required by plants to complete their normal life cycles which includes C, H, O, P, K, N, S, Ca, Fe, Mg, Mn, Cu, B, Zn, Co, Mo, Cl, and Na.

estuary–A seaward end or the widened funnel-shaped tidal mouth of a river valley where fresh water comes into contact with seawater and where tidal effects are evident; e.g., a tidal river, or a partially enclosed coastal body of water where the tide meets the current of a stream.

eubacteria–Prokaryotes other than **archaebacteria**.

euic–High level of bases in soil material, specified at family level of classification.

eukaryote–Cellular organisms having a membrane-bound nucleus within which the genome of the cell is stored as chromosomes composed of DNA; includes algae, fungi, protozoa, plants, and animals.

eutrophic–Having concentrations of nutrients optimal, or nearly so, for plant, animal, or microbial growth. (Said of nutrient or soil solutions and bodies of water.) The term literally means "self-feeding."

evaporites–Residue of salts (including gypsum and all more soluble species) precipitated by evaporation.

evapotranspiration–The combined loss of water from a given area, and during a specified period of time, by evaporation from the soil surface and by transpiration from plants.

exchangeable anion–A negatively charged ion held on or near the surface of a solid particle by a positive surface charge and which may be easily replaced by other negatively charged ions (e.g. with a Cl$^-$ salt).

exchangeable bases–Charge sites on the surface of soil particles that can be readily replaced with a salt solution. In most soils, Ca^{2+}, Mg^{2+}, K^+ and Na^+ predominate. Historically, these are called bases because they are cations of strong bases. Many soil chemists object to this term because these cations are not bases by any modern definition of the term. See also **base saturation** and **exchangeable cation**.

exchangeable cation–A positively charged ion held on or near the surface of a solid particle by a negative surface and which may be replaced by other positively charged ions in the soil solution. Usually expressed in centimoles or millimoles of charge per kilogram.

exchangeable cation percentage–(no longer preferred in SSSA publications) The extent to which the adsorption complex of a soil is occupied by a particular cation. It is expressed as:

$$ECP = \frac{\text{exchangeable cation (cm kg}^{-1}\text{ soil)}}{\text{cation exchange capacity (cmol kg}^{-1}\text{ soil)}} \times 100$$

exchangeable nutrient–A plant nutrient that is held by the adsorption complex of the soil and is easily exchanged with the anion or cation of neutral salt solutions.

exchangeable sodium fraction–The fraction of the cation exchange capacity of a soil occupied by sodium ions.

exchangeable sodium percentage (ESP)–Exchangeable sodium fraction expressed as a percentage.

exchangeable sodium ratio (ESR)–The ratio of exchangeable sodium to all other exchangeable cations.

exoenzyme–Enzymes that are excreted by organisms into the surrounding environment and carry out their metabolic or catabolic activity in that location.

experimental plot–The smallest area unit in field studies that receives an experimental treatment.

extractable soil nutrient–The quantity of a nutrient removed from the soil by a specific soil test procedure.

extragrade–(i) A taxonomic class at the subgroup level of soil taxonomy having properties that are not characteristic of any class in a higher category (any order, suborder or great group) and that do not indicate transition to any other known kind of soil. (ii) A soil that is a member of one such subgroup. See also **intergrade**.

exudate, root–Low molecular weight metabolites that enter the soil from plant roots.

F

fabric–The physical constitution of soil material as expressed by the spatial arrangement of the solid particles and associated voids.

facies–The sum of all primary lithologic and paleontologic characteristics of sediments or sedimentary rock that are used to infer its origin and environment; the general nature of appearance of sediments or sedimentary rock produced under a given set of conditions; a distinctive group of characteristics that distinguish one group from another within a stratigraphic unit; e.g. contrasting river-channel facies and overbank-flood-plain facies in alluvial valley fills.

facultative organism–An organism that can carry out both options of a mutually exclusive process (e.g., aerobic and anaerobic metabolism). May also be used in reference to other processes, such as photosynthesis (e.g., a facultative photosynthetic organism is one that can use either light or the oxidation of organic or inorganic compounds as a source of energy).

faecal (fecal) pellets–Rounded and subrounded aggregates of fecal material produced by the soil fauna.

fall cone–A variety of cone penetrometer which utilizes dropping weights to provide known increments of force applied to the cone, resulting in measured increments of soil penetration.

fallow–See **tillage**, *fallow*.

family, soil–In soil classification one of the categories intermediate between the subgroup and the soil series. Families provide groupings of soils with ranges in texture, mineralogy, temperature, and thickness. See also **classification**, **soil**.

fan, alluvial–A generic term for constructional landforms that are built of stratified **alluvium** with or without debris-flow deposits and that occur on the pediment slope, downslope from their source of alluvium.

fault–A fracture or fracture zone of the earth with displacement along one side in respect to the other.

fen–A peat accumulating wetland that receives some drainage from surrounding mineral soils and usually supports marsh-like vegetation. These areas are richer in nutrients and less acidic than bogs. The soils under fens are **peat (Histosols)** if the **fen** has been present for a while. See also **bog, pocosin, swamp**, and **wetland**.

fermentation–The metabolic process in which an organic compound serves as both an electron donor and the final electron acceptor.

ferran–A **cutan** composed of iron oxides, hydroxides, or oxyhydroxides.

ferri-argillan–A **cutan** consisting of a mixture of clay minerals and iron oxides, hydroxides, or oxyhydroxides.

ferrihydrite–$Fe_5O_7(OH)\cdot4H_2O$. A dark reddish-brown, poorly crystalline iron oxide mineral that forms in wet soils. Occurs in **concretions** and **placic horizons** and often can be found in ditches and pipes that drain wet soils.

Ferrods–**Spodosols** that have more than six times as much free iron (elemental) than organic carbon in the spodic horizon. Ferrods are rarely saturated with water or do not have characteristics associated with wetness. (A suborder in the U.S. system of soil taxonomy.)

ferrolysis–Clay destruction process involving disintegration and solution in water based upon the alternate reduction and oxidation of iron.

fertigation–Application of plant nutrients in irrigation water.

fertility, soil–The relative ability of a soil to supply the nutrients essential to plant growth.

fertilization, foliar–Application of a dilute solution of liquid fertilizers to plant foliage.

fertilizer–Any organic or inorganic material of natural or synthetic origin (other than liming materials) that is added to a soil to supply one or more plant nutrients essential to the growth of plants.

  acid-forming–Fertilizer that, after application to and reaction with soil, increases residual acidity and decreases soil pH.

  blended–A mechanical mixture of different fertilizer materials.

  bulk-blended–A physical mixture of dry granular fertilizer materials to produce specific fertilizer ratios and grades. Individual granules in the bulk blended fertilizer do not have the same ratio and content of plant food as does the mixture as a whole.

  complete–A chemical compound or a blend of compounds that contains significant quantities of all three primary nutrients, N, P, and K. It may contain other plant nutrients.

  compound–A fertilizer formulated with two or more plant nutrients.

  controlled-release–A fertilizer term used interchangeably with delayed release, slow release, controlled availability, slow acting, and metered release to designate a controlled dissolution of fertilizer at a lower rate than conventional water-soluble fertilizers. Controlled-release properties may result from coatings on water-soluble fertilizers or from low dissolution and/or mineralization rates of fertilizer materials in soil.

  granular–Fertilizer particles sized between an upper and lower limit or between two screen sizes, usually within the range of 1 to 4 mm and often more closely sized. The desired size may be obtained by agglomerating smaller particles, crushing and screening larger particles, controlling size in crystallization processes, or prilling.

  injected–Placement of fertilizer into the soil either through use of pressure or nonpressure systems.

inorganic–A fertilizer material in which carbon is not an essential component of its basic chemical structure.

liquid–Fertilizer wholly or partially in solution that can be handled as a liquid, including clear liquids and liquids containing solids in suspension.

mixed–Two or more fertilizer materials blended or granulated together into individual mixes. The term includes dry mix powders, granulated, clear liquid, suspension, and slurry mixtures.

organic–A material containing carbon and one or more plant nutrients in addition to hydrogen and/or oxygen.

pop-up–Fertilizer placed in small amounts in direct contact with the seed.

salt index–The ratio of the decrease in osmotic potential of a solution containing a fertilizer compound or mixture to that produced by the same weight of $NaNO_3 \times 100$.

sidedressed–A fertilizer application usually banded to the side of crop rows after plant emergence.

slow-release–See **fertilizer**, *controlled-release*.

starter–A fertilizer applied in relatively small amounts with or near the seed usually during planting for the purpose of accelerating early growth of the crop plants.

suspension–A fluid fertilizer containing dissolved and undissolved plant nutrients. The undissolved plant nutrients are kept in suspension with a suspending agent, usually a swelling type clay. The suspension must be flowable enough to be mixed, pumped, agitated, and applied to the soil in a homogeneous mixture.

top-dressed–A non-incorporated surface application of fertilizer to a soil after the crop has been established.

fertilizer analysis–The percent composition of a fertilizer as determined in a laboratory and expressed as total N, available phosphoric acid (P_2O_5) equivalent, and water-soluble potash (K_2O) equivalent.

fertilizer fixation–See **fixation**.

fertilizer grade–The guaranteed minimum analysis in percent of the major plant nutrient elements contained in a fertilizer material or in a mixed fertilizer. The analysis is usually designated as $N–P_2O_5–K_2O$; but it may be N–P–K where permitted or required as specified by state law. Grades must be expressed in percent N–P–K for SSSA publications (oxide values may be included in parentheses). See also **fertilizer analysis**.

fertilizer ratio–The relative proportions of primary nutrients in a **fertilizer grade** divided by the highest common denominator for that grade, e.g., grades 10–6–4 and 20–12–8 have a ratio 5–3–2.

fertilizer recommendation–See **soil test interpretation.**

fertilizer requirement–The quantity of certain plant nutrients needed to increase nutrient availability in the soil with the objective of increasing plant growth to a designated level.

fibric material–Organic soil material that contains ¾ or more recognizable fibers (after rubbing between fingers) of undecomposed plant remains. Bulk density is usually very low and water holding capacity very high.

Fibrists–Histosols that have a high content of undecomposed plant fibers and a **bulk density** less than about 0.1 g cm^{-3}. Fibrists are saturated with water for periods long enough to limit their use for most crops unless they are artificially drained. (A suborder in the U.S. system of soil taxonomy.)

field capacity, in situ (field water capacity)–The content of water, on a mass or volume basis, remaining in a soil 2 or 3 days after having been wetted with water and after free drainage is negligible. See also **available water**.

field strip cropping–See **tillage**, *strip cropping*.

fifteen-atmosphere percentage–(no longer used in SSSA publications) The percentage of water contained in a soil that has been saturated, subjected to, and is in equilibrium with, an applied pressure of 15 atm. Approximately the same as **fifteen-bar percentage**. See also **soil water**.

fifteen-bar percentage–(no longer used in SSSA publications) The percentage of water contained in a soil that has been saturated, subjected to, and is in equilibrium with, an applied pressure of 15 bars. Approximately the same as the **fifteen-atmosphere percentage**. See also **soil water**.

film water–A thin layer of water, in close proximity to soil-particle surfaces, that varies in thickness from 1 or 2 to perhaps 100 or more molecular layers.

fine sand–(i) A soil separate. See also **soil separates**. (ii) A soil textural class. See also **soil texture**.

fine sandy loam–A soil textual class. See also **soil texture**.

fine texture–(i) A broad group of textures consisting of or containing large quantities of the fine fractions, particularly of silt and clay. (Includes all sandy clay, silty clay, and clay textural classes). (ii) When used in reference to family particle-size classes in U.S. soil taxonomy, is specifically defined as having 35 to 60 percent clay. See also **soil texture**.

finger–A vertically elongated path of preferential water flow in soil. See also **preferential flow**.

fire, ground–(forestry) A fire that consumes all organic material of the forest floor and also burns into the underlying soil itself, as, for example, a peat fire. Differentiated from a surface fire on the basis of vulnerability to wind; in a surface fire the flames are visible and burning is accelerated by wind, whereas, in a ground fire, wind is generally not a serious factor.

firm–A soil consistence term. See also **consistence**.

firming–See **tillage**, *firming*.

first bottom–The lowest and most frequently flooded part of the flood plain of a stream.

fixation–The process by which available plant nutrients are rendered less available or unavailable in the soil. Not to be confused with **dinitrogen fixation**.

flaggy–Containing appreciable quantities of **flagstones**. See also **rock fragments**.

flagstone–A relatively thin, flat rock fragment, from 150 to 380 mm on the long axis. See also **rock fragments**.

flat planting–See **tillage**, *flat planting*.

flexible cropping–A strategy of growing adapted crops with cropping and fallow decisions at each prospective date of planting based on available water in the soil plus expected growing season precipitation and without regard to a predetermined rigidly adhered to cropping sequence.

flocculation–The coagulation of colloidal soil particles due to the ions in solution. In most soils the clays and humic substances remain flocculated due to the presence of doubly and triply charged cations.

flood plain–The nearly level plain that borders a stream and is subject to inundation under flood-stage conditions unless protected artificially. It is usually a constructional landform built of sediment deposited during overflow and lateral migration of the stream.

flooding–Accumulation of large amounts of runoff on the landscape as a result of rainfall in excess of the soil's ability to drain water from the landscape before extensive inundation and ponding occurs. See also **irrigation**.

flow velocity (of water in soil)–The volume of water transported per unit of time and per unit of cross-sectional area normal to the direction of water flow.

flowtill–A supraglacial **till** that is modified and transported by mass flow.

flume–See **irrigation**, *flume*.

fluorescent antibody–An antiserum conjugated with a fluorescent dye, (e.g., fluorescein or rhodamine). Fluorescent-labeled antiserum can be used to stain burred slides or other preparations and visualize the specific microorganism (antigen) of interest by fluorescence microscopy. See also **immunofluorescence**.

Fluvents–**Entisols** that form in recent loamy or clayey **alluvial** deposits, are usually stratified, and have an organic carbon content that decreases irregularly with depth. **Fluvents** are not saturated with water for periods long enough to limit their use for most crops. (A suborder in the U.S. system of soil taxonomy.)

fluvioglacial–See **glaciofluvial deposits**.

flux–The time rate of transport of a quantity (e.g., mass or volume of fluid, electromagnetic energy, number of particles, or energy) across a given area. See also **flux density**.

flux density–The time rate of transport of a quantity (e.g., mass or volume of fluid, electromagnetic energy, number of particles, or energy) per unit area perpendicular to the direction of flow.

f_{oc}–Fraction of organic carbon in a soil

foliar diagnosis–An estimation of plant mineral nutrient status from the chemical composition of selected plant parts, and the color and growth characteristics of the plant foliage.

Folists–**Histosols** that have an accumulation of organic soil materials mainly as forest litter that is <1 m deep to rock or to fragmental materials with interstices filled with organic materials. Folists are not saturated with water for periods long enough to limit their use if cropped. (A suborder in the U.S. system of soil taxonomy.)

footslope–The hillslope position that forms the inner, gently inclined surface at the base of a slope. In profile, footslopes are commonly concave and are situated between the **backslope** and a **toeslope**.

forest floor–All organic matter generated by forest vegetation, including litter and unincorporated humus, on the mineral soil surface.

forest productivity–The capacity of a forest to produce specific products (i.e. biomass, lumber) over time as influenced by the interaction of vegetative manipulation and abiotic factors (i.e. soil, climate, physiography). **Net primary productivity (NPP)** provides the fundamental measure of forest productivity.

When measured at the point of foliar carrying capacity for all potential flora, NPP is a measure of potential site productivity. Rate of product growth, an economic component, is occasionally used as a partial measure of forest productivity.

fracture–A planar void between **aggregates**.

fragile land–Land that is sensitive to degradation when disturbed; such as with highly erodible soils, soils where salts can and do accumulate, and soils at high elevations.

fragipan–A natural subsurface horizon with very low organic matter, high **bulk density** and/or high mechanical strength relative to overlying and underlying horizons; has hard or very hard **consistence** (seemingly cemented) when dry, but showing a moderate to weak brittleness when moist. The layer typically has **redoximorphic features**, is slowly or very slowly permeable to water, is considered to be root restricting, and usually has few to many bleached, roughly vertical planes which are faces of coarse or very coarse polyhedrons or prisms.

free iron oxides–A general term for those iron oxides that can be reduced and dissolved by a dithionite treatment. Generally includes **goethite**, **hematite**, **ferrihydrite**, **lepidocrocite**, and **maghemite**, but not **magnetite**. See also **iron oxides**.

friable–A consistency term pertaining to the ease of crumbling of soils. See also **consistence**.

friction cone penetrometer–A **cone penetrometer** with the additional capacity of measuring the local side-friction component of **penetration resistance**. The resistance to penetration developed by the friction sleeve equals the vertical force applied to the sleeve divided by its surface area. See also **cone penetrometer**, **penetration resistance**, and **cone index**.

frigid–A soil temperature regime that has mean annual soil temperatures of >0° C but <8° C, >5° C difference between mean summer and mean winter soil temperatures at 50 cm below the surface, and warm summer temperatures. Isofrigid is the same except the summer and winter temperatures differ by <5° C.

fritted trace elements–Sintered silicates having total guaranteed analyses of micronutrients with controlled, relatively slow, release characteristics.

frost heaving–Lifting or lateral movement of soil as caused by freezing processes in association with the formation of ice lenses or ice needles.

frost, concrete–Ice in the soil in such quantity as to constitute a virtually solid block.

frost, honeycomb–Ice in the soil in insufficient quantity to be continuous, thus giving the soil an open, porous structure permitting the ready entrance of water.

fulvic acid–The pigmented organic material that remains in solution after removal of **humic acid** by acidification. It is separated from the fulvic acid fraction by adsorption on a hydrophobic resin at low pH values. See also **soil organic matter**.

fulvic acid fraction–Fraction of soil organic matter that is soluble in both alkali and dilute acid.

functional nutrient–Chemical elements that function in plant metabolism whether or not their action is specific.

fungistat–A compound that inhibits or prevents fungal growth.

furrow–See **tillage**, *furrow*; **irrigation**, *furrow*.

furrow mulching–See **erosion**, *furrow mulching*.

G

gas pressure (external) potential–See **air pressure** in **Table 5**.

genetic–Resulting from, or produced by, soil-forming processes; for example, a genetic soil profile or a genetic horizon.

geographic information system (GIS)–A method of overlaying large volumes of spatial data of different kinds. The data are referenced to a set of geographical coordinates and encoded in a form suitable for handling by a digital computer. Different data planes can be overlain, statistically analyzed, and used to make estimates of soil and land suitabilities.

geological erosion–See **erosion**, *geological erosion*.

geomorphic surface–A mappable area of the earth's surface that has a common history; the area is of similar age and is formed by a set of processes during an episode of landscape evolution. A geomorphic surface can be erosional, constructional or both. The surface shape can be planar, concave, convex, or any combination of these.

geomorphology–The science that studies the evolution of the earth's surface. The science of landforms. The systematic examination of landforms and their interpretation as records of geologic history.

Gibbs free energy (G)–The thermodynamic potential for a system whose independent variables are the absolute temperature, applied pressure, mass variables, and other independent, extensive variables. The change in Gibbs free energy, as a system passes reversibly from one state to another at constant temperature and pressure, is a measure of the work available in that change of state.

gibbsite–$Al(OH)_3$. A mineral with a platy habit that occurs in highly weathered soils and in laterite. Also, may be prominent in the subsoil and **saprolite** of soils formed on crystalline rock high in feldspar.

gilgai–The microrelief of small basins and knolls or valleys and ridges on a soil surface produced by expansion and contraction during wetting and drying (usually in regions with distinct, seasonal, precipitation patterns) of clayey soils that contain **smectite**. See also **microrelief**.

glacial drift–A general term applied to all mineral material transported by a glacier and deposited directly by or from the ice, or by running water emanating from a glacier. Drift includes unstratified material (till) that forms **moraines**, and stratified **glaciofluvial deposits** that form **outwash** plains, **eskers**, **kames**, **varves**, and **glaciolacustrine sediments**.

glacial till–See **till** (i).

glacial soil–A soil derived from **glacial drift**. (Not used in current U.S. system of soil taxonomy.)

glaciers–Large masses of ice that formed, in part, on land by the compaction and recrystallization of snow. They may be moving downslope or outward in all directions because of the stress of their own weight or they may be retreating or be stagnant.

glaciofluvial deposits–Material moved by **glaciers** and subsequently sorted and deposited by streams flowing from the melting ice. The deposits are stratified and may occur in the form of **outwash** plains, **deltas**, **kames**, **eskers**, and kame terraces. See also **glacial drift** and **till** (i).

glaciolacustrine deposits–Material ranging from fine clay to sand derived from **glaciers** and deposited in glacial lakes by water originating mainly from the melting of glacial ice. Many are bedded or laminated with **varves**.

glaebule–A three-dimensional unit within the **s-matrix** of the soil material. Its morphology is incompatible with its present occurrence being within a single void in the present soil material. It is recognized as a unit either because of a greater concentration of some constituent and/or a difference in fabric compared with the enclosing soil material, or because it has a distinct boundary with the enclosing soil material.

glauconite–A Fe-rich dioctahedral mica with tetrahedral Al (or Fe^{3+}) usually greater than 0.2 atoms per formula unit and octahedral R^{3+} correspondingly greater than 1.2 atoms. A generalized formula is $K(R_{1.33}{}^{3+}R_{0.67}{}^{2+})(Si_{3.67}Al_{0.33})O_{10}(OH)_2$ with $Fe^{3+}>>Al$ and $Mg>Fe(II)$ (unless altered). Further characteristics are d(060) >0.151 nm and (usually) broader infrared spectra than celadonite. Mixtures containing an iron-rich mica as a major component can be called glauconitic.

Gley soil–Soil developed under conditions of poor drainage resulting in reduction of iron and other elements and in gray colors and mottles. (Not used in current U.S. system of soil taxonomy.)

gleyed–A soil condition resulting from prolonged soil saturation, which is manifested by the presence of bluish or greenish colors through the soil mass or in mottles (spots or streaks) among the colors. Gleying occurs under reducing conditions, by which iron is reduced predominantly to the ferrous state.

gleyzation–A soil-forming process resulting in the development of gley soils. (Not used in current U.S. system of soil taxonomy.)

goethite–FeOOH. A yellow-brown iron oxide mineral. Goethite occurs in almost every soil type and climatic region, and is responsible for the yellowish-brown color in many soils and weathered materials.

grain cutan–**Cutan** associated with the surfaces of skeleton grains or other discrete units such as **nodules**, **concretions**, etc.

grain density–See **particle density**.

granular soil structure–A shape of soil structure. See also **soil structure** and **soil structure shapes**.

granulation–The process of producing granular materials.

granule–A natural soil aggregate or ped of relatively low porosity. See also **soil structure** and **soil structure shapes**.

grassed waterway–See **erosion**, *grassed waterway*.

gravelly–Containing appreciable amounts of pebbles. See also **rock fragments**.

gravitational potential–See **soil water**.

gravitational water–Water which moves into, through, or out of the soil under the influence of gravity. See also **soil water**, *soil water potential*.

gravitropism–The natural tendency for biological organisms or specific cells or organs of an organism to respond to the stimulus of gravity.

gravity flow–See **irrigation**, *gravity flow*.

gravity irrigation–See **irrigation**, *gravity*.

Gray-Brown Podzolic soil–A **zonal** great soil group consisting of soils with a thin, moderately dark A1 (A) horizon and with a grayish-brown A2 (E) horizon underlain by a B horizon containing a high percentage of bases and an appreciable quantity of **illuviated** silicate clay; formed on relatively young land surfaces, mostly glacial deposits, from material relatively rich in calcium, under deciduous forests in humid temperate regions. (Not used in current U.S. system of soil taxonomy.)

Gray Desert soil–A term used in Russia, and frequently in the USA, synonymously with Desert soil. (Not used in current U.S. system of soil taxonomy.)

great soil group–One of the categories in the system of soil classification that has been used in the USA for many years. Great groups place soils according to soil moisture and temperature, base saturation status, and expression of horizons. See also **classification, soil**.

green manure–Plant material incorporated into soil while green or at maturity, for soil improvement.

green manure crop–Any crop grown for the purpose of being turned under while green or soon after maturity for soil improvement.

greenhouse effect–The absorption of solar radiant energy by the earth's surface and its release as heat into the atmosphere; longer infrared heat waves are absorbed by the air, principally by carbon dioxide and water vapor, thus, the atmosphere traps heat much as does the glass in a greenhouse.

gross duty of water–See **irrigation**, *gross duty of water.*

gross primary productivity (GPP)–Total carbon assimilation by plants. GPP = **NPP** + respiration losses. See also **net primary productivity**.

ground data–Supporting data collected on the ground, and information derived there from, as an aid to the interpretation of remotely recorded surveys, such as airborne imagery, etc. Generally, this should be performed concurrently with the airborne surveys. Data as to weather, soils, and vegetation types and conditions are typical.

ground moraine–An extensive layer of **till**, having an uneven or undulating surface; a deposit of rock and mineral debris dragged along, in, on, or beneath a **glacier** and emplaced by processes including basal lodgement and release from downwasting stagnant ice by **ablation**.

ground water–That portion of the water below the surface of the ground at a pressure equal to or greater than atmospheric. See also **water table.**

Ground-Water Laterite soil–A great soil group of the **intrazonal** order and **hydromorphic** suborder, consisting of soils characterized by **hardpans** or concretional horizons rich in iron and aluminum (and sometimes manganese) that have formed immediately above the water table. (Not used in current U.S. system of soil taxonomy.)

Ground-Water Podzol soil–A great soil group of the **intrazonal** order and **hydromorphic** suborder, consisting of soils with an organic mat on the surface over a very thin layer of acid humus material underlain by a whitish-gray leached layer, which may be as much as 2 or 3 feet in thickness, and is underlain by a brown, or very dark-brown, cemented **hardpan** layer; formed under various types of forest vegetation in cool to tropical, humid climates under conditions of poor drainage. (Not used in current U.S. system of soil taxonomy.)

guano–The decomposed dried excrement of birds and bats, used for fertilizer.

guess row–See **tillage**, *guess row.*

gullied land–Areas where all **diagnostic soil horizons** have been removed by water, resulting in a network of V-shaped or U-shaped channels. Some areas resemble miniature badlands. Generally, gullies are so deep that extensive reshaping is necessary for most uses.

gully–See **erosion**, *gully*.

gypsan–A cutan composed of gypsum.

gypsic horizon–A mineral soil horizon of secondary $CaSO_4$ enrichment that is >15 cm thick, has at least 50 g kg^{-1} more gypsum than the C horizon, and in which the product of the thickness in centimeters and the amount of $CaSO_4$ is equal to or greater than 1500g kg^{-1}.

Gypsids–**Aridisols** which have a **gypsic** or **petrogypsic horizon** that has its upper boundary within 100 cm of the soil surface and lack a petrocalcic horizon overlying any of these horizons. (A suborder in the U.S. system of soil taxonomy.)

gypsum–$CaSO_4 \cdot 2H_2O$. The common name for calcium sulfate, used to supply calcium to ameliorate soils with a high exchangeable sodium fraction.

gypsum requirement–The quantity of gypsum or its equivalent required to reduce the exchangeable sodium fraction of a given amount of soil to an acceptable level where dispersion of soil colloids does not take place.

gyttja–Sedimentary peat consisting mainly of plant and animal residues precipitated from standing water.

H

habitat–The place where a given organism lives.

Half-Bog soil–A great soil group, of the **intrazonal** order and **hydromorphic** suborder consisting of soil with dark-brown or black peaty material over grayish and rust mottled mineral soil; formed under conditions of poor drainage under forest, sedge, or grass vegetation in cool to tropical humid climates. (Not used in current U.S. system of soil taxonomy.)

halloysite–A member of the **kaolin** subgroup of **clay minerals**. It is similar to **kaolinite** in structure and composition except that hydrated varieties occur that have interlayer water molecules. Halloysite usually occurs as tubular or spheroidal particles and is most common in soils formed from volcanic ash. See also **Appendix I, Table A3**.

halomorphic soil–A suborder of the intrazonal soil order, consisting of **saline** and **sodic** soils formed under imperfect drainage in arid regions and including the great soil groups **Solonchak** or Saline soils, **Solonetz** soils, and Soloth soils. (Not used in current U.S. system of soil taxonomy.)

hardpan–A soil layer with physical characteristics that limit root penetration and restrict water movement.

hardsetting soil–Soils that, following wetting, exhibit transient but only slowly reversible cementation and/or induration throughout significant fractions of the profile restrictive to seed emergence and root penetration (Australian).

harrowing–See **tillage**, *harrowing*.

harvest index–The quantity of harvestable biomass per unit total biomass produced. If used in relation to nutrients it would be the quantity of biomass produced per unit input of plant nutrient.

head land–See **tillage**, *turnrow*.

headcut–See **erosion**, *headcut*.

heat flux–See **soil heat flux density**.

heat of immersion–The heat evolved on immersing a soil, at a known initial water content (usually oven dry) in a large volume of water.

heavy metals–Those metals which have densities >5.0 Mg m^{-3}. In soils these include the elements Cd, Co, Cr, Cu, Fe, Hg, Mn, Mo, Ni, Pb, and Zn.

heavy soil–(colloquial) A soil with a high content of the fine separates, particularly clay, or one with a high drawbar pull and hence difficult to cultivate, especially when wet. See also **fine texture**.

hematite–Fe_2O_3. A red iron oxide mineral that contributes red color to many soils.

hemic material–Organic soil material at an intermediate degree of decomposition that contains 1/6 to 3/4 recognizable fibers (after rubbing) of undecomposed plant remains. Bulk density is usually very low, and water holding capacity very high.

Hemists–Histosols that have an intermediate degree of plant fiber decomposition and a bulk density between about 0.1 and 0.2 g cm^{-3}. **Hemists** are saturated with water for periods long enough to limit their use for most crops unless they are artificially drained. (A suborder in the U.S. system of soil taxonomy.)

heterotroph–An organism able to derive carbon and energy for growth and cell synthesis by utilizing organic compounds.

heterotrophic nitrification–Biochemical oxidation of ammonium and/or organic nitrogen to nitrate and nitrite by heterotrophic microorganisms. See also **nitrification**.

hill–See **tillage**, *hill*.

histic epipedon–A thin organic soil horizon that is saturated with water at some period of the year unless artificially drained and that is at or near the surface of a mineral soil. The **histic epipedon** has a maximum thickness depending on the kind of materials in the horizon and the lower limit of organic carbon is the upper limit for the **mollic epipedon**.

Histosols–Organic soils that have organic soil materials in more than half of the upper 80 cm, or that are of any thickness if overlying rock or fragmental materials that have interstices filled with organic soil materials. (An order in the U.S. system of soil taxonomy.) See also **peat**.

hoe–See **tillage**, *hoe*.

horizon–See **soil horizon**.

hue–A measure of the chromatic composition of light that reaches the eye; one of the three variables of color. See also **Munsell color system**, **chroma**, and **value, color**.

humic acid–The dark-colored organic material that can be extracted from soil with dilute alkali and other reagents and that is precipitated by acidification to pH 1 to 2.

Humic Gley soil–Soil of the intrazonal order and **hydromorphic** suborder that includes Wisenboden and related soils, such as **Half-Bog soils**, which have a thin muck or peat O2 (Oi) horizon and an A1 (A) horizon. Developed in wet meadow and in forested swamps. (Not used in current U.S. system of soil taxonomy.)

humic substances–A series of relatively high-molecular-weight, yellow to black colored organic substances formed by secondary synthesis reactions in soils. The term is used in a generic sense to describe the colored material or its fractions obtained on the basis of solubility characteristics. These materials are distinctive to soil environments in that they are dissimilar to the biopolymers of microorganisms and higher plants (including lignin). See also **humic acid**, **fulvic acid**, and **humin**.

humification–The process whereby the carbon of organic residues is transformed and converted to humic substances through biochemical and abiotic processes.

humin–The fraction of the soil organic matter that cannot be extracted from soil with dilute alkali.

Humods–Spodosols that have accumulated organic carbon and aluminum, but not iron, in the upper part of the spodic horizon. **Humods** are rarely saturated with water or do not have characteristics associated with wetness. (A suborder in the U.S. system of soil taxonomy.)

Humox–Oxisols that are moist all or most of the time and that have a high organic carbon content within the upper 1 m. **Humox** have a mean annual soil temperature of <22° C and a base saturation within the **oxic horizon** of <35%, measured at pH 7. (A suborder in the U.S. system of soil taxonomy.)

Humults–Ultisols that have a high content of organic carbon. **Humults** are not saturated with water for periods long enough to limit their use for most crops. (A suborder in the U.S. system of soil taxonomy.)

humus–Total of the organic compounds in soil exclusive of undecayed plant and animal tissues, their "partial decomposition" products, and the soil biomass. The term is often used synonymously with **soil organic matter**.

humus form–A group of soil horizons located at or near the surface of a pedon, which have formed from organic residues, either separate from or intermixed with, mineral material.

hybridization–The binding or annealing of two, complementary, single strands of nucleic acid.

hydrated lime–A liming material composed mainly of calcium and magnesium hydroxides that reacts quickly to neutralize acid soils.

hydraulic conductivity–See **soil water**, *hydraulic conductivity*.

hydraulic gradient (soil water)–A vector (macroscopic) point function that is equal to the decrease in the hydraulic head per unit distance through the soil in the direction of the greatest rate of decrease. In isotropic soils, this will be in the direction of the water flux.

hydraulic head–See **soil water**.

hydric soils–Soils that are wet long enough to periodically produce anaerobic conditions, thereby influencing the growth of plants.

hydrodynamic dispersion–The process wherein the solute concentration in flowing solution changes in response to the interaction of solution movement with the pore geometry of the soil, a behavior with similarity to diffusion but only taking place when solution movement occurs.

hydrodynamic dispersion coefficient–The coefficient in the solute convection equations that accounts for **hydrodynamic dispersion**, it is usually determined by solving an inverse problem.

hydrogen bond–An intramolecular chemical bond between a hydrogen atom of one molecule and a highly electronegative atom (e.g. O, N) of another molecule.

hydrogenic soil–Soil developed under the influence of water standing within the profile for considerable periods; formed mainly in cold, humid regions.

hydrologic cycle–The fate of water from the time of precipitation until the water has been returned to the atmosphere by evaporation and is again ready to be precipitated.

hydrology–The science dealing with the distribution and movement of water.

   agrohydrology–The science dealing with the distribution and movement of rainfall and/or irrigation water and soil solution to and from the root zone in agricultural land, and with the distribution and movement of irrigation and surface water in conveyance systems on agricultural land.

   ground-water hydrology–The science dealing with the movement of the soil solution in the saturated zone of the soil profile.

   soil hydrology–The science dealing with the distribution and movement of the soil solution in the soil profile.

   surface hydrology–The science dealing with the distribution and conveyance of water on the soil surface.

   wetland hydrology–The science dealing with water depth, flow patterns and duration, and frequency of flooding that define and delineate wetlands.

hydromorphic soils–A suborder of intrazonal soils, consisting of seven great soil groups, all formed under conditions of poor drainage in marshes, swamps, seepage areas, or flats. (Not used in current U.S. system of soil taxonomy.)

hydrophilic–Molecules and surfaces that have a strong affinity for water molecules.

hydrophobic–Molecules and surfaces that have little or no affinity for water molecules. Hydrophobic substances have more affinity for other hydrophobic substances than for water.

hydrophobic soils–Soils that are water repellent, often due to dense fungal mycelial mats or hydrophobic substances vaporized and reprecipitated during fire.

hydroseeding–See **erosion**, *hydroseeding*.

hydrous mica–A better term might be **illite**.

hydroxy-aluminum interlayers–Polymers of general composition $[Al(OH)_{3-x}]_m^{mx+}$ which are adsorbed on interlayer cation exchange sites. Although not exchangeable by unbuffered salt solutions, they are responsible for a considerable portion of the titratable acidity (and pH-dependent charge) in soils.

hydroxy-interlayered vermiculite–A **vermiculite** with partially filled interlayers of hydroxy-aluminum groups. It is normally dioctahderal in both the interlayer and the octahedral sheet of the vermiculite layer. It is common in the coarse clay fraction of acid surface soil horizons. It has intermediate cation exchange properties between vermiculite and chlorite. Synonyms are "chlorite-vermiculite intergrade"; "vermiculite-chlorite intergrade." See also **hydroxy-aluminum interlayers**.

hygroscopic coefficient–(no longer used in SSSA publications) The weight percentage of water held by, or remaining in, the soil (i) after the soil has been air-dried, or (ii) after the soil has reached equilibrium with an unspecified environment of high relative humidity, usually near saturation, or with a specified relative humidity at a specified temperature.

hygroscopic water–(no longer used in SSSA publications) Water adsorbed by a dry soil from an atmosphere of high relative humidity, water remaining in the soil after "air-drying," or water held by the soil when it is in equilibrium with an atmosphere of a specified relative humidity at a specified temperature, usually 98% relative humidity at 25°C.

hyperthermic–A soil temperature regime that has mean annual soil temperatures of 22° C or more and >5° C difference between mean summer and mean winter soil temperatures at 50 cm below the surface. Isohyperthermic is the same except the summer and winter temperatures differ by <5° C.

hypha (pl. hyphae)–Filament of fungal cells. Many hyphal filaments (hyphae) constitute a **mycelium**. Bacteria of the order *Actinomycetales* also produce branched mycelium.

hypo-coating–Pedofeatures related to natural surfaces in soils that occur superposed to the adjoining groundmass rather than on the surface. Similar to **neocutan**.

hypoxic–Insufficient availability of oxygen in an environment to support aerobic respiration.

hysteresis–A nonunique relationship between two variables, wherein the curves depend on the sequences or starting point used to observe the variables. Examples include the relationships: (i) between soil-water content and soil-water matric potential, (ii) between solution concentration and adsorbed quantity of chemical species, and (iii) between soil volume and water content for swelling and shrinking soils.

I

igneous rock–Rock formed from the cooling and solidification of magma, and that has not been changed appreciably by weathering since its formation. See also **metamorphic rock**.

illite–(i) As a general term, refers to either a discrete non-expansible mica of detrital or authigenic origin or to the micaceous component of interstratified systems, as in illite-smectite. If used to refer to the *species*, it should meet the following requirements: a) The micaceous layers ideally are non-expansible; b) the octahedral sheet is dioctahedral and aluminous; c) the interlayer cation is primarily potassium; and (4) the composition deviates from that of muscovite in two main ways: 1) A phengitic component is present in which substitution of R^{2+} cations for octahedral Al is balanced by addition of tetrahedral Si beyond the ideal Si:Al ratio of 3:1 for **muscovite**. This substitution gives the octahedral sheet an overall negative charge of about 0.2 to 0.3 per formula unit. 2) Interlayer vacancies or water molecules amounting to about 0.2 to 0.4 atoms per formula unit are compensated by additional tetrahedral Si cations beyond those required by the phengitic component. Where reference is made to the *species* illite, a clear statement should be made to that effect in order to avoid confusion with the *general* usage. (ii) In soil taxonomy, the presence of a 1 nm x-ray diffraction peak and ≥4% K_2O is used to denote the presence of illite.

illuvial horizon–A soil layer or horizon in which material carried from an overlying layer has been precipitated from solution or deposited from suspension. The layer of accumulation. See also **eluvial horizon**.

illuviation–The process of deposition of soil material removed from one horizon to another in the soil; usually from an upper to a lower horizon in the soil profile. See also **eluviation**.

illuviation cutan–See **clay films.**

imagery–General term for base map or reference map materials.

Imhoff cone–A graduated volumetric cone used for determining settleable solids in liquid suspensions.

immobilization–The conversion of an element from the inorganic to the organic form in microbial or plant tissues.

immunofluorescence–Fluorescence resulting from a reaction between a substance and a specific antibody that is bound to a fluorescent dye.

imogolite–A poorly crystalline aluminosilicate mineral with an ideal composition SiO_2 Al_2O_3 $2.5H_2O(+)$. It appears as threads consisting of assemblies of a tube unit with inner and outer diameters of 1.0 and 2.0 nm, respectively. **Imogolite** is commonly found in association with **allophane**, and is similar to **allophane** in chemical properties. **Imogolite** is mostly found in soils derived from volcanic ash, and in weathered pumices and **Spodosols**.

impeded drainage–A condition which hinders the movement of water through soils under the influence of gravity.

impervious–Resistant to penetration by fluids or by roots.

Inceptisols–Mineral soils that have one or more pedogenic horizons in which mineral materials other than carbonates or amorphous silica have been altered or removed but not accumulated to a significant degree. Under certain conditions, Inceptisols may have an **ochric**, **umbric**, **histic**, **plaggen** or **mollic epipedon**. Water is available to plants more than half of the year or more than 90 consecutive days during a warm season. (An order in the U.S. system of soil taxonomy.)

inclusion–One or more polypedons or parts of polypedons within a delineation of a map unit, not identified by the map unit name; i.e., is not one of the named component soils or named miscellaneous area components. Such soils or areas are either too small to be delineated separately without creating excessive map or legend detail, or occur too erratically to be considered a component, or are not identified by practical mapping methods. See also **component soil, map unit**.

incorporation–See **tillage**, *incorporation*.

indicator plants–Plants characteristically associated with specific soil or site conditions, such as soil acidity, alkalinity, wetness, or a chemical element.

indigenous–Native to an area.

indurated–A very strongly **cemented** soil horizon. See also **consistence**.

infiltrability–The flux (or rate) of water infiltration into soil when water at atmospheric pressure is maintained on the atmosphere-soil boundary, with the flow direction being one-dimensionally downward.

infiltration capacity–See **infiltration flux.**

infiltration flux (or rate)–The volume of water entering a specified cross-sectional area of soil per unit time $[L\ t^{-1}]$.

infiltration, cumulative–See **cumulative infiltration.**

infiltration–The entry of water into soil.

infiltrometer–A device for measuring the volume or flux (or rate) of liquid (usually water) entry downward into the soil.

infrared (abbr. IR)–Pertaining to or designating the portion of the electromagnetic spectrum with wavelengths just beyond the red end of the visible spectrum in the wavelength interval from about 0.75 μm to 1 mm.

infrared, far–A term for the longer wavelengths of the infrared region, from 25 μm to 1 mm, the generally accepted shorter wavelength limit of the microwave part of the EM spectrum. This is severely limited in terrestrial use, as the atmosphere transmits very little radiation between 25 μm and the millimeter regions.

infrared, middle–A term for the mid section of the infrared region of the electromagnetic spectrum with wavelengths from around 2 or 3 μm (varying with the author), to around 25 μm. This is the region commonly referred to when discussing the infrared spectra of chemical compounds, organic or inorganic, and minerals.

infrared, near–The preferred term for the shorter wavelengths in the infrared region extending from about 0.75 μm (visible red), to around 2 or 3 μm (varying with the author). The longer wavelength end grades into the middle infrared. The term really emphasizes the radiation reflected from plant materials, which peaks around 0.85 μm. It is also called solar infrared, as it is only available for use during the daylight hours.

inner sphere adsorption–Adsorption of ions that occurs with the elimination of water of hydration in the space between the adsorbed ion and the surface. The force of retention of ions involves both ionic and covalent bonding. Strong adsorption of anions and cations at variable charge sites in **organic matter**, oxides, and **phyllosilicate** edges involves inner sphere adsorption.

inoculate–To treat, usually seeds, with microorganisms to create a favorable response. Most often refers to the treatment of legume seeds with *Rhizobium* or *Bradyrhizobium* to stimulate dinitrogen fixation but also refers to the introduction of microbial cultures into sterile growth medium.

in-row subsoiling–See **tillage**, *in-row subsoiling*.

integrated drainage–a general term for a drainage pattern in which stream systems have developed to the point where all parts of the landscape drain into some part of a stream system, the initial or original surfaces have essentially disappeared and the region drains to a common base level.

interception–See **precipitation interception**.

interflow–That portion of rainfall that infiltrates into the soil and moves laterally through the upper soil horizons until intercepted by a stream channel or until it returns to the surface at some point downslope from its point of infiltration.

interfluve–A landform composed of the relatively undissected upland or ridge between two adjacent valleys or drainageways.

intergrade–(i) A taxonomic class at the subgroup level of soil taxonomy having properties typical of the great group of which it is a member and that are characteristic of some class in a higher category (any order, suborder or great group) and indicates a transition to that kind of soil. (ii) A soil that is a member of one such subgroup. See also **extragrade**.

interlayer–See **phyllosilicate mineral terminology**.

intermittent stream–A stream, or reach of a stream, that does not flow year-round and that flows only when a) it receives baseflow solely during wet periods, or b) it receives ground-water discharge or protracted contributions from melting snow or other erratic surface and shallow subsurface sources.

internal friction–The portion of the shearing strength of a soil indicated by the term $\sigma \tan\theta$ in Coulomb's equation $\tau = c + \sigma \tan\theta$, where τ is shear stress, σ is normal stress, c is cohesion, and θ is friction angle. It is usually considered to be due to the interlocking of soil grains and the resistance to sliding between the grains.

interstratification–Mixing of different kinds of silicate layers along the c-direction in a given stack. Interstratification may be regular or random. In regular interstratification, the stacking of the component layers follows a periodic succession. In random interstratification, the distribution of the different layers lacks periodicity and is controlled only by the proportions of the various layers.

intrazonal soils–(i) One of the three orders in soil classification. (ii) A soil with more or less well developed soil characteristics that reflect the dominating influence of some local factor of relief, parent material, or age, over the normal effect of climate and vegetation. (Not used in current U.S. system of soil taxonomy.)

intrinsic permeability–The property of a porous material that expresses the ease with which gases or liquids flow through it. Often symbolized by $k = K\eta/\rho g$, where K is the Darcy hydraulic conductivity, η is the fluid viscosity, ρ is the fluid density, and g is the acceleration of gravity. Dimensionally, k is an area $[L^2]$. See also **permeability** and **soil water**.

inverse problem–Determining the properties of a system from its response to a known stimulus.

inversion–See **tillage**, *inversion*.

ion activity–Single ion activity is calculated by multiplying the concentration by the activity coefficient, usually calculated using the extended Debye-Hückel equation or the Davies equation. Numerically, it approaches the value of the ionic concentration at infinite dilution. See also **activity (chemical)**.

ion selective electrode–An electrochemical sensor, the potential of which (in conjunction with a suitable **reference electrode**) depends on the logarithm of the activity of a given ion in aqueous solution (e.g. pH, copper, nitrate, and sodium electrodes).

ion selectivity–(i) The relative adsorption of an ion by the solid phase in relation to the adsorption of other ions. (ii) The relative absorption of an ion by a root in relation to absorption of other ions.

ionic strength–A parameter that estimates the interaction between ions in solution. It is calculated as one-half the sum of the products of ionic concentration and the square of ionic charge for all the charged species in a solution. It is needed for calculation of single **ion activity**.

ions–Atoms, groups of atoms, or compounds, which are electrically charged as a result of the loss of electrons (cations) or the gain of electrons (anions).

iron oxides–Group name for the oxides and hydroxides of iron. Includes the minerals **goethite**, **hematite**, **lepidocrocite**, **ferrihydrite**, **maghemite**, and **magnetite**. Sometimes referred to as "sesquioxides," or "iron hydrous oxides."

iron pan–A **hardpan** in which iron oxide is the principal cementing agent. Also spelled: ironpan. See also **plinthite**.

ironstone–An in-place concentration of iron oxides that is at least weakly cemented.

irrigable area–See **irrigation**, *irrigable area*.

irrigation–The intentional application of water to the soil, usually for the purpose of crop production. Related terms include:

advance time–The time it takes the first water applied to a dry irrigation furrow to travel the length of the furrow.

alternate set irrigation–A method of managing irrigation whereby, at every other irrigation, alternate furrows are irrigated, or sprinklers are placed midway between their locations during the previous irrigation.

alternate side irrigation–The practice of furrow irrigating one side of a crop row (for row crops or orchards) and then, at about half the irrigation time, irrigating the other side.

border dikes–Earth ridges built to guide or hold irrigation water within prescribed limits in a field; a small levee.

border ditch irrigation–A ditch used as a border of an irrigated strip or plot, water being spread from one or both sides of the ditch along its entire length.

border-strip irrigation–The water is applied at the upper end of a strip with earth borders to confine the water to the strip.

center-pivot irrigation–Automated sprinkler irrigation achieved by automatically rotating the sprinkler pipe or boom, supplying water to the sprinkler heads or nozzles, as a radius from the center of the field to be irrigated. Water is delivered to the center or pivot point of the system. The pipe is supported above the crop by towers at fixed spacings and propelled by pneumatic, mechanical, hydraulic, or electric power on wheels or skids in fixed circular paths at uniform angular speeds. Water is applied at a uniform rate by progressive increase of nozzle size from the pivot to the end of the line. The depth of water applied is determined by the rate of travel of the system. Single units are ordinarily about 1250 to 1300 feet long (381 to 397 m) and irrigate approximately a 130-acre (52.7-ha) circular area.

check-basin irrigation–The water is applied rapidly to relatively level plots surrounded by levees. The basin is a small check.

check irrigation–Modification of a border strip with small earth ridges or checks constructed at intervals to retain water as the water flows down the strip.

conjunctive water use–The joining together of two sources of irrigation water, such as groundwater and surface water, to serve a particular piece of land.

consumptive irrigation requirement–The centimeters per hectare of irrigation water, exclusive of precipitation, stored soil moisture, or ground water, needed consumptively for crop production.

continuous delivery–A system by which an irrigator receives his allotted quantity of water at a continuous rate throughout the irrigation season.

contour ditch–Irrigation ditch laid out approximately on the contour.

contour flooding–Method of irrigating by flooding from contour ditches.

contour-furrow irrigation–Applying irrigation water in furrows that run across the slope with a forward grade in the furrows.

contour-level irrigation–Irrigation of areas bounded by small contour levels; cross levels are completely flooded.

controlled drainage–(irrigation) Regulation of the water table to maintain the water level at a depth favorable for optimum crop growth.

conveyance loss–Loss of water from delivery systems during conveyance, including operational losses and losses due to seepage, evaporation, and transpiration by plants growing in or near the channel.

corrugate irrigation–The water is applied to small, closely spaced furrows called corrugates, frequently in grain and forage crops, to confine the flow of irrigation water to one direction.

cutback irrigation–Water applied in furrow irrigation at a faster rate at the beginning of the irrigation period and then reduced or cutback to a lesser rate, usually one-half the initial rate or that amount to balance with the intake rate.

demand system of irrigation–System of irrigation water delivery where each irrigator may request irrigation water in the amount needed and at the time desired.

discharge curve–(i) Rating curve showing the relation between stage and rate of flow of a stream. (ii) Curve showing the relation of discharge of a pump and the speed, power, and head.

drainage curves–Design curves giving prescribed rates of surface runoff for different levels of crop production, and which may vary according to size of drainage area.

drip irrigation–Irrigation whereby water is slowly applied to the soil surface through small emitters having low-discharge orifices.

dynamic head–The total of the following factors: a) the total static head, b) friction head in the discharge pipeline, c) head losses in fittings, elbows, and valves, and d) pressure required to operate lateral lines.

flood irrigation–Irrigation in which the water is released from field ditches and allowed to flood over the land.

flume–(i) Open conduit for conveying water across obstructions. (ii) An entire canal elevated above natural ground. An aqueduct. (iii) A specially calibrated structure for measuring open channel flows.

furrow irrigation–Irrigation in which the water is applied between crop rows in furrows made by tillage implements.

gravity–Irrigation in which the water is not pumped but flows and is distributed by gravity.

gravity flow–Water flow which is not pumped but flows due to the acceleration forces of gravity. Used in irrigation, drainage, inlets, and outlets.

gravity sprinkler–A sprinkler irrigation system in which gravity furnishes the desired head.

gross duty of water–The irrigation water diverted at the intake of a canal system, usually expressed in depth on the irrigable area under the system; diversion requirement. See also **irrigation**, *net duty of water*.

gross irrigation water requirement–The net water requirement plus distribution and application losses in operating the system.

irrigable area–Area capable of being irrigated, principally as regards to availability of water, suitable soils, and topography of land.

irrigation application efficiency–Percentage of irrigation water applied to an area that is stored in the soil for crop use.

irrigation canal–A permanent irrigation canal constructed to convey water from the source of supply to one or more farms.

irrigation check–Small dike or dam used in the furrow alongside an irrigation border to make the water spread evenly across the border.

irrigation efficiency–Variously defined, including: (i)The ratio of the water actually consumed by crops on an irrigated area to the amount of water applied to the area; (ii) the ratio of water infiltrated to total water applied; (iii) the ratio of water profile storage increase to total water applied.

irrigation frequency–Time interval between irrigations.

irrigation hose–A closed conduit for supplying water to moving irrigation systems, flexible when subjected to normal operating pressure and may be collapsible to a flat cross section when purged of water.

irrigation lateral–A branch of a main canal conveying water to a farm ditch; sometimes used in reference to farm ditches.

irrigation methods–The methods and/or manner in which water is intentionally applied to an area.

irrigation period–The number of hours or days that it takes to apply one irrigation to a given design area during the peak consumptive-use period of the crop being irrigated.

irrigation tailwater recovery system–A water runoff collection and storage system to provide a constant quantity of water back to the initial system or to another field. Water is applied to the rows at the same rate for the entire irrigation period. Advance time should equal irrigation recession time as nearly as possible. Recession time is usually one-fourth of the entire irrigation period.

irrigation set–The area irrigated at one time within a field.

lagtime–(flood irrigation) The period between the time that the irrigation stream is turned off at the upper end of an irrigated area and the time that water disappears from the surface at the point or points of application.

lath box–Preferred term is spile. See **irrigation**, *spile*.

length of run–Distance water must run in furrows or between borders over the surface of a field from one head ditch to another, or to the end of the field.

limited irrigation–Management of irrigation applications to apply less than enough water to satisfy the soil water deficiency in the entire root zone. Sometimes called "deficit" or "stress irrigation".

line source–Continuous source of water emitted along a line.

manifold–Pipeline that supplies water to the laterals.

micro irrigation–See **irrigation**, *trickle*.

miner's inch–The rate of discharge through an orifice 1-inch square under a specified head. An old term used in the western USA, now seldom used except where irrigation or mining water rights are so specified. The equivalent flow in cubic feet per second is fixed by state statute. One miner's inch is equivalent to 0.025 cubic foot per second in Arizona, California, Montana, and Oregon; 0.020 cubic foot per second in British Columbia.

net duty of water–The amount of water delivered to the land to produce a crop, measured at the point of delivery to the field. See also **irrigation**, *gross duty of water*.

percent area wetted–Area wetted by irrigation as a percentage of the total crop area.

preplant irrigation–Irrigation applied prior to seeding. Sometimes called "preirrigation".

rotation irrigation–A system by which irrigators receive an allotted quantity of water, not a continuous rate, but at stated intervals; for example, a number of irrigators receiving water from a lateral may agree to rotate the water, each taking the entire flow in turn for a limited period.

screen–(i) (wells) A manufactured well casing with precisely dimensioned and shaped openings. (Compare with perforated casing.) (ii) (canals) A device used to clean surface water of debris, such as revolving screens or turbulent fountain screens.

siphon tubes–Small curved pipes, typically 0.5 to 4.0 inches (1.3 to 10.2 cm) in diameter, that deliver water over the side of a head ditch or lateral to furrows, corrugations, or borders.

spile–A wooden box that is placed in a ditch bank to transfer water from an irrigation ditch to the field to be irrigated. This is the preferred term instead of *lath box*.

spray irrigation–The application of water by a small spray or mist to the soil surface, where travel through the air becomes instrumental in the distribution of water.

sprinkler–The water is broadcast over the entire soil surface through spray nozzles or high volume guns utilizing a pressurized system.

sprinkler irrigation systems terms

> *boom*–An elevated, cantilevered sprinkler(s) mounted on a central stand. The sprinkler boom rotates about a central pivot.

> *center pivot*–An automated irrigation system consisting of a sprinkler line rotating about a pivot point and supported by a number of self-propelled towers. The water is supplied at the pivot point and flows outward through the line supplying the individual outlets. See also **irrigation, center pivot.**

> *corner pivot*–An additional span or other equipment attached to the end of a center pivot irrigation system that allows the overall radius to increase or decrease in relation to the field boundaries.

> *lateral move*–An automated irrigation machine consisting of a sprinkler line supported by a number of self-propelled towers. The entire unit moves in a generally straight path and irrigates a basically rectangular area. Sometimes called a "linear move".

> *microirrigation*–The frequent application of small quantities of water and drops, tiny, streams, or miniature spray through emitters or applicators placed along a water delivery line. Microirrigation encompasses a number of methods or concepts such as bubbler, drip, trickle, mist, or spray.

> *mist irrigation*–A method of microirrigation in which water is applied in very small droplets.

> *nozzle*–Discharge opening or orifice of a sprinkler head used to control the volume of discharge, distribution pattern, and droplet size.

> *permanent*–Underground piping with risers and sprinklers.

> *portable (hand move)*–Sprinkler system which is moved by uncoupling and picking up the pipes manually, requiring no special tools.

> *reel and gun irrigation (traveling gun)*–A form of irrigation utilizing a single large rotating gun mounted on a frame to deliver water in a circle.

Water is supplied from flexible hosing and the gun can either be pulled manually to new stations for each irrigation set or gradually pulled by cable on a timer.

side-move sprinkler–A sprinkler system with the supply pipe supported on carriages and towing small diameter trailing pipelines, each fitted with several sprinkler heads.

side-roll sprinkler–The supply pipe is usually mounted on wheels with the pipe as the axle and where the system is moved across the field by rotating the pipeline by engine power.

solid set–System which covers the complete field with pipes and sprinklers in such a manner that all the field can be irrigated without moving any of the system.

sprinkler distribution pattern–Water depth-distance relationship measured from a single sprinkler head.

towed sprinkler–System where lateral lines are mounted on wheels, sleds, or skids, and are pulled or towed in a direction approximately parallel to the lateral.

subbing–(i) The process of a crop obtaining water directly, from a shallow water table. (ii) (colloquial) The horizontal movement of water from an irrigation furrow to the row bed.

subirrigation–The water is applied in open ditches or tile lines until the water table is raised sufficiently to supply water to the rooting depth of the crop.

supplemental irrigation–Irrigation to insure increased crop production in areas where rainfall normally supplies most of the moisture needed.

surface irrigation–Irrigation where the soil surface is used as a conduit, as in furrow and border irrigation as opposed to sprinkler irrigation or subirrigation.

surge irrigation–A surface irrigation technique wherein flow is applied to furrows (or less commonly, borders) intermittently during a single irrigation set.

tailwater–(i) (hydraulics) Water, in a river or channel, immediately down-stream from a structure. (ii) (irrigation) Water that reaches the lower end of a field.

tailwater recovery–The process of collecting irrigation water runoff for reuse in the system.

trickle–Water applied slowly through a system of low volume hoses or tubes, above or below the soil surface, under low pressure from small openings.

trickle irrigation–A method of microirrigation wherein water is applied to the soil surface as drops or small streams through emitters. (Preferred term is **drip irrigation**.)

emitter–A small microirrigation dispensing device designed to dissipate pressure and discharge a small uniform flow or trickle of water at a constant discharge, which does not vary significantly because of minor differences in pressure head. Also called a "dripper" or "trickler".

compensating emitter–Designed to discharge water at a constant rate over a wide range of lateral line pressures.

continuous flushing emitter–Designed to continuously permit passage of large solid particles while operating at a trickle or drip flow thus reducing filter fineness requirements.

flushing emitter–Designed to have a flushing flow of water to clear the discharge opening every time the system is turned on.

line-source emitter–Water is discharged from closely spaced perforations, emitters, or a porous wall along the tubing.

long path emitter–Employs a long capillary-sized tube or channel to dissipate pressure.

multi-outlet emitter–Supplies water to 2 or more points through small diameter auxiliary tubing.

orifice emitter–Employs a series of orifices to dissipate pressure.

porous trickle tubing–Tubing with a uniformly porous wall. The pores are small and ooze water under pressure.

subsurface drip irrigation–Application of water below the soil surface through emitters, with discharge rates generally in the same range as drip irrigation. This method of water application is different from and not to be confused with subirrigation where the root zone is irrigated by water table control.

vortex emitter–Employs a vortex effect to dissipate pressure.

wild-flooding–The water is released at high points in the field and distribution is uncontrolled.

irrigation-induced erosion–See **erosion**, *irrigation-induced erosion*.

isoelectric point–The activity of potential determining ion in a solution in equilibrium with a variable charge surface whose net electrical charge is zero. For soils it refers to the pH of the isoelectric point of pH dependent charge materials. It applies only to single components, not mixtures.

isomorphous substitution–The replacement of one atom by another of similar size in a crystal structure without disrupting or seriously changing the structure. When a substituting cation is of a smaller valence than the cation it is replacing, there is a net negative charge on the structure.

isotopically exchangeable ion–An ion, bonded to a solid surface that can exchange with similar isotopically labeled ions in solution in a specified period of time.

J

jarosite–$KFe_3(OH)_6(SO_4)_2$. A pale yellow potassium iron sulfate mineral.

joint planes–Planar voids that traverse the soil material in some fairly regular pattern, such as parallel or subparallel sets.

K

K_2O–Potassium oxide, designation on the fertilizer label that denotes the percentage of available potassium reported as K_2O. See also **potash**.

kame–A low mound, knob, hummock, or short irregular ridge, composed of stratified sand and gravel deposited by a subglacial stream as a **fan** or **delta** at the margin of a melting **glacier**; by a supraglacial stream in a low place or hole on the surface of the glacier; or as a ponded deposit on the surface or at the margin of stagnant ice.

kandic horizon–Subsoil **diagnostic horizon** having a clay increase relative to overlying horizons and has low activity clays i.e., <160 cmol$_c$ kg^{-1} clay.

kaolin–A subgroup name of aluminum silicates with a 1:1 layer structure. **Kaolinite** is the most common clay mineral in the subgroup. Also, a soft, usually white, rock composed largely of kaolinite. See also **Appendix I, Table A3**.

kaolinite–A clay mineral of the **kaolin** subgroup. It has a 1:1 layer structure composed of shared sheets of Si-O tetrahedrons and Al-(O,OH) octahedrons with very little **isomorphous substitution**. See also **Appendix I, Table A3**.

karst–Topography with sinkholes, caves, and underground drainage that is formed in limestone, gypsum, or other rocks by dissolution.

K$_d$–See **distribution coefficient, K$_d$**.

K$_{oc}$–The **distribution coefficient, K$_d$**, calculated on the basis of organic carbon content. $K_{oc}=K_d/f_{oc}$ where f_{oc} is the fraction of organic carbon.

K$_{ow}$–The octanol-water partition coefficient. The ratio of the concentration of an organic compound in octanol and in water after equilibration of the two phases. Can be used to estimate the value of K_{oc} for some organic compounds.

kriging–A method based on the theory of regionalized variables for predicting without bias and minimum variance the spatial distribution of earth components, including soil properties.

krotovina–Irregular tubular streaks within one layer of material transported from another layer by filling of tunnels made by burrowing animals with material from outside the layer in which they are found.

K-selected–In ecological theory, that group of microorganisms in soil living at or near the carrying capacity of the soil environment. Analogous to **autochthonous** microorganisms.

L

labile–Readily transformed by microorganisms or readily available to plants.

labile pool–The sum of an element in the soil solution and the amount of that element readily solubilized or exchanged when the soil is equilibrated with a salt solution.

lacustrine deposit–**Clastic** sediments and chemical precipitates deposited in lakes.

lacustrine soil–Soil formed on or from **lacustrine deposits**.

lagtime–See **irrigation**, *lagtime*.

land–(i) The entire complex of surface and near surface attributes of the solid portions of the surface of the earth, which are significant to human activities; water bodies occurring within land masses are included in some land classification systems. (ii) (economics) One of the major factors of production that is supplied by nature and includes all natural resources in their original state, such as mineral deposits, wildlife, timber, fish, water, coal, and the fertility of the soil.

land capability–The suitability of land for use without permanent damage. Land capability, as ordinarily used in the USA, is an expression of the effect of physical land conditions, including climate, on the total suitability for use, without damage, for crops that require regular tillage, for grazing, for woodland, and for wildlife. Land capability involves consideration of the risks of land damage from erosion and other causes and the difficulties in land use owing to physical land characteristics, including climate.

land capability class–One of the eight classes of land in the land capability classification of the U.S. Natural Resource Conservation Service; distinguished according to the risk of land damage or the difficulty of land use; they include:

Land suitable for cultivation and other uses.

Class I–Soils that have few limitations restricting their use.

Class II–Soils that have some limitations, reducing the choice of plants or requiring moderate conservation practices.

Class III–Soils that have severe limitations that reduce the choice of plants or require special conservation practices, or both.

Class IV–Soils that have very severe limitations that restrict the choice of plants, require very careful management or both.

Land generally not suitable for cultivation (without major treatment).

Class V–Soils that have little or no erosion hazard, but that have other limitations, impractical to remove, that limit their use largely to pasture, range, woodland, or wildlife food and cover.

Class VI–Soils that have severe limitations that make them generally unsuited for cultivation and limit their use largely to pasture or range, woodland, or wildlife food and cover.

Class VII–Soils that have very severe limitations that make them unsuited to cultivation and that restricts their use largely to grazing, woodland, or wildlife.

Class VIII–Soils and landforms that preclude their use for commercial plant production and restrict their use to recreation, wildlife, water supply, or aesthetic purposes.

land capability subclass–Groups of capability units within classes of the land capability classification that have the same kinds of dominant limitations for agricultural use as a result of soil and climate. Some soils are subject to erosion if they are not protected, while others are naturally wet and must be drained if crops are to be grown. Some soils are shallow or droughty or have other soil deficiencies. Still other soils occur in areas where climate limits their use. The four kinds of limitations recognized at the subclass level are: risks of erosion, designated by the symbol (e); wetness, drainage, or overflow (w); other root zone limitations (s); and climatic limitations (c). The subclass provides the map user information about both the degree and kind of limitation. Capability Class I has no subclasses.

land evaluation–The process of assessment of land performance when the land is used for specific purposes.

land farming–A process of bioremediation or biodegradation in which wastes are incorporated into soil and allowed to decompose via naturally occurring microbial activity.

landform–Any physical, recognizable form or feature on the earth's surface, having a characteristic shape, and produced by natural causes; it includes a wide range in size such as a **shrub-coppice dune** than can be several meters across vs. a seif dune which can be up to 100 km long. Landforms provide an empirical description of similar portions of the earth's surface.

landforming–See **tillage**, *landforming*.

land planing–See **tillage**, *land planing*.

landscape–A collection of related **landforms**; usually the land surface which the eye can comprehend in a single view.

landslide–A general term for a mass movement **landform** and a process characterized by moderately rapid to rapid (greater than 30 cm per year) downslope transport, by means of gravitational stresses, of a mass of rock and **regolith** that may or may not be water saturated.

lapilli–Non or slightly vesicular **pyroclastics**, 2.0 to 76 mm in at least one dimension, with an apparent specific gravity of 2.0 or more.

Lateritic soil–A suborder of **zonal soils** formed in warm, temperate, and tropical regions and including the following great soils groups: Yellow Podzolic, Red Podzolic, Yellowish-Brown Lateritic, and Lateritic. (Not used in current U.S. system of soil taxonomy.)

lath box–See **irrigation**, *spile*.

Latosol–A suborder of **zonal soils** including soils formed under forested, tropical, humid conditions and characterized by low silica-sesquioxide ratios of the clay fractions, low base-exchange capacity, low activity of the clay, low content of most primary minerals, low content of soluble constituents, a high degree of aggregate stability, and usually having a red color. (Not used in current U.S. system of soil taxonomy.)

lattice–A regular geometric arrangement of points in a plane or in space. Lattice is used to represent the distribution of repeating atoms or groups of atoms in a crystalline substance. A lattice is a mathematical concept. Atomic substitutions take place in a *structure* and not in a lattice. See also **phyllosilicate mineral terminology**.

lattice energy–The energy required to separate the ions of a crystal to an infinite distance.

lava flow–A solidified body of rock formed from the lateral, surficial outpouring of molten lava from a vent or fissure, often lobate in form.

law of diminishing returns–When other factors in production do not change, successive increases in the input of one factor will not proportionately increase product yield.

law of the minimum–See **Liebig's law**

layer charge–Magnitude of charge per formula unit of a clay which is balanced by ions of opposite charge external to the unit layer. See also **phyllosilicate mineral terminology**.

layer silicate minerals–Synonymous with the term **phyllosilicates**. See also **Appendix I**, **Table A3**.

layer–See **phyllosilicate mineral terminology**.

leachate–Liquids that have percolated through a soil and that contain substances in solution or suspension.

leaching–The removal of soluble materials from one zone in soil to another via water movement in the profile. See also **eluviation.**

leaching fraction–The fraction of infiltrated irrigation water that percolates below the root zone.

leaching requirement–The leaching fraction necessary to keep soil salinity, chloride, or sodium (the choice being that which is most demanding) from exceeding a tolerance level of the crop in question. It applies to steady-state or long-term average conditions.

lectins–Plant proteins that have a high affinity for specific sugar residues.

leghemoglobin–An iron-containing, red pigment(s) produced in root nodules during the symbiotic association between *Bradyrhizobium* or *Rhizobium* and legumes. The pigment resembles but is not identical to mammalian hemoglobin.

length of run–See **irrigation**, *length of run*.

lepidocrocite–FeOOH An orange iron oxide mineral that is found in **mottles** and **concretions** of wet soils.

Liebig's law–The growth and reproduction of an organism is dependent on the nutrient substance that is available in minimum quantity.

lift–See **tillage**, *lift*.

light soil–(colloquial) A coarse-textured soil; a soil with a low drawbar pull and hence easy to cultivate. See also **coarse textured** and **soil texture**. Contrast to **heavy soil**.

lime, agricultural–A soil amendment containing calcium carbonate, magnesium carbonate and other materials, used to neutralize soil acidity and furnish calcium and magnesium for plant growth. Classification including **calcium carbonate equivalent** and limits in lime particle size is usually prescribed by law or regulation.

lime concretion–An aggregate of precipitated calcium carbonate, or of other material cemented by precipitated calcium carbonate.

lime-pan–A hardened layer cemented by calcium carbonate. Better term may be **caliche**.

lime requirement–The amount of liming material as calcium carbonate equivalent required to change a volume of soil to a specified state with respect to pH or soluble Al content.

limited irrigation See **irrigation**, *limited irrigation*.

limnic material–One of the common components of organic soils and includes both organic and inorganic materials that were either (i) deposited in water by precipitation or through the action of aquatic organisms, or (ii) derived from underwater and floating aquatic plants and aquatic animals.

line source–See **irrigation**, *line source*.

liquid limit–The minimum mass water content at which a small sample of soil will barely flow under a standard treatment. Synonymous with "upper plastic limit." See also **Atterberg limits**, **consistency**, **plastic limit**, and **plasticity number**.

lister planting–See **tillage**, *lister planting*.

listing–See **tillage**, *listing*.

lithic contact–A boundary between soil and continuous, coherent, underlying material. The underlying material must be sufficiently coherent to make hand-digging with a spade impractical. If a single mineral, its hardness is 3 (Mohs scale), and gravel size chunks that do not disperse with 15 hours shaking in water or sodium hexametaphosphate solution.

lithiophorite–(Al,Li)MnO$_2$(OH)$_2$ A black manganese oxide that is common in iron-manganese **nodules** of acid soils. It has a layer structure.

lithorelict–A micromorphological feature derived from the parent rock that can be recognized by its rock structure and fabric.

lithosequence–A group of related soils that differ, one from the other, in certain properties primarily as a result of differences in the parent material as a soil-forming factor.

Lithosols–A great soil group of **azonal soils** characterized by an incomplete solum or no clearly expressed soil morphology and consisting of freshly and imperfectly weathered rock or rock fragments. (Not used in current U.S. system of soil taxonomy.)

litter–The surface layer of the forest floor which is not in an advanced stage of decomposition, usually consisting of freshly fallen leaves, needles, twigs, stems, bark, and fruits.

loam–A soil textural class. See also **soil texture**.

loamy–(i) Texture group consisting of coarse sandy loam, sandy loam, fine sandy loam, very fine sandy loam, loam, silt loam, silt, clay loam, sandy clay loam, and silty clay loam soil textures. See also **soil texture**. (ii) Family particle-size class for soils with textures finer than very fine sandy loam but <35% clay and <35% rock fragments in upper subsoil horizons.

loamy coarse sand–A soil textural class. See also **soil texture**.

loamy fine sand–A soil textural class. See also **soil texture**.

loamy sand–A soil textural class. See also **soil texture**.

loamy very fine sand–A soil textural class. See also **soil texture**.

lodgement till–A **basal till** characterized by compact, fissile or platy structure and containing coarse fragments oriented with their long axes generally parallel to the direction of ice movement.

loess–Material transported and deposited by wind and consisting of predominantly silt-sized particles.

loose–A soil consistence term. See also **consistence**.

loosening–See **tillage**, *loosening.*

lower plastic limit–See **plastic limit**.

lowland vs. upland soils–Terms commonly used to denote landscape positions that are subject to flooding or that are deliberately flooded for rice production vs. those that are not.

luxury uptake–The absorption of nutrients by plants in excess of that quantity needed for optimum growth. Luxury concentrations during early growth may be utilized in later growth.

lysimeter–(i) A device for measuring percolation and leaching losses from a column of soil under controlled conditions. (ii) A device for measuring gains (irrigation, precipitation, and condensation) and losses (evapotranspiration) by a column of soil.

M

macronutrient–A plant nutrient found at relatively high concentrations (>500 mg kg^{-1}) in plants. Usually refers to N, P, and K, but may include Ca, Mg, and S.

macropore–See **pore-size classification**.

macropore flow–The tendency for water applied to the soil surface at rates exceeding the upper limit of unsaturated hydraulic conductivity, to move into the soil profile mainly via saturated flow through macropores, thereby bypassing micropores and rapidly transporting any solutes to the lower soil profile. See also **preferential flow**.

made land–Areas filled with earth, or with earth and trash mixed, usually by or under the control of man. See also **miscellaneous areas**.

maghemite–Fe_2O_3 A dark reddish-brown, magnetic iron oxide mineral chemically similar to **hematite**, but structurally similar to **magnetite**. Often found in well-drained, highly weathered soils of tropical regions.

magnetite–Fe_3O_4 A black, magnetic iron oxide mineral usually inherited from igneous rocks. Often found in soils as black magnetic sand grains.

maintenance application–Application of fertilizer materials in amounts and at intervals to maintain available soil nutrients at levels necessary to produce a desired yield.

mangan–A **cutan** composed of manganese oxide or hydroxide.

manganese oxides–A group term for oxides of manganese. They are typically black and frequently occur in soils as **nodules** and coatings on ped faces usually in association with iron oxides. **Birnessite** and **lithiophorite** are common manganese oxide minerals in soils.

manifold–See **irrigation**, *manifold*.

manure–The excreta of animals, with or without an admixture of bedding or litter, fresh or at various stages of further decomposition or composting. In some countries may denote any fertilizer material.

map, large-scale–A map having a scale of 1:100 000 or larger.

map, medium-scale–A map having a scale from 1:100 000, exclusive, to 1:1 000 000, inclusive.

map, small-scale–A map having a scale smaller than 1: 1 000 000.

map unit, soil–(i) A conceptual group of one to many delineations identified by the same name in a soil survey that represent similar landscape areas comprised of either: (1) the same kind of component soil, plus inclusions, or (2) two or more kinds of component soils, plus inclusions, or (3) component soils and miscellaneous area, plus inclusions, or (4) two or more kinds of component soils that may or may not occur together in various delineations but all have similar, special use and management, plus inclusions, or (5) a miscellaneous area and included soils. (ii) A loose synonym for a delineation. See also **delineation, component soil, inclusion, soil consociation, soil complex, soil association, undifferentiated group, miscellaneous areas**.

marl–Soft and unconsolidated calcium carbonate, usually mixed with varying amounts of clay or other impurities.

marsh–A wet area, periodically inundated with standing or slow moving water, that has grassy or herbaceous vegetation and often little peat accumulation; the water may be salt, brackish or fresh. Sometimes called wet prairies. See also **swamp, tidal flats**, and **wetland**.

mass flow (nutrient)–The movement of solutes associated with net movement of water.

mass movement–Dislodgement and downslope transport of soil and rock material as a unit under direct gravitational stress. The process includes slow displacements such as **creep** and **solifluction**, and rapid movements such as **landslides**, rock slides, and falls, earthflows, debris flows, and **avalanches**. Agents of fluid transport (water, ice, air) may play an important, if subordinate role in the process.

matric potential–See **soil water**, *soil water potential*.

matric suction–(Term is no longer used in SSSA publications.) The preferred term is **matric potential**. See also **soil water**, *soil water potential*.

mature soil–A soil with well-developed soil horizons produced by the natural processes of soil formation and essentially in equilibrium with its present environment.

meander land–Unsurveyed land along a lake shore or stream border that has developed by the receding of the shore line or of the stream since the last cadastral survey of the area. See also **miscellaneous areas**.

mechanical analysis–See **particle size analysis** and **particle size distribution**.

mechanical weathering–The process of weathering by which frost action, salt-crystal growth, absorption of water, and other physical processes break down a rock into smaller fragments; no chemical change is involved.

medium-textured–Texture group consisting of very fine sandy loam, loam, silt loam, and silt textures. See also **soil texture**.

mesic–A soil temperature regime that has mean annual soil temperatures of 8° C or more but <15° C, and >5° C difference between mean summer and mean winter soil temperatures at 50 cm below the surface. Isomesic is the same except the summer and winter temperatures differ by <5° C.

mesobiota–See **mesofauna**.

mesofauna–Nematodes, oligochaete worms, smaller insect larvae, and microarthropods.

mesophile–See **mesophilic organism**.

mesophilic organism–An organism whose optimum temperature for growth falls in an intermediate range of approximately 15 to 35° C. Synonymous with "mesophile."

metamorphic rock–Rock derived from preexisting rocks that have been altered physically, chemically, and/or mineralogically as a result of natural geological processes, principally heat and pressure, originating within the earth. The preexisting rocks may have been **igneous, sedimentary**, or another form of metamorphic rock.

mica–A layer-structured aluminosilicate mineral group of the 2:1 type that is characterized by its non expandability and high layer charge, which is usually satisfied by potassium. The major types are **muscovite, biotite**, and phlogopite. See also **Appendix I, Table A3**.

microaerophile–An organism that requires a low concentration of oxygen for growth. Sometimes used to indicate an organism that will carry out its metabolic activities under aerobic conditions but that will grow much better under anaerobic conditions.

microbial biomass–(i) The total mass of living microorganisms in a given volume or mass of soil. (ii) The total weight of all microorganisms in a particular environment.

microbial population–The sum of living microorganisms in a given volume or mass of soil.

microbiota–Microflora and protozoa.

microclimate–(i) The climatic condition of a small area resulting from the modification of the general climatic conditions by local differences in elevation or exposure or other local phenomena. (ii) The sequence of atmospheric changes within a very small region.

microfauna–Protozoa, nematodes, and arthropods of microscopic size.

microflora–Bacteria (including actinomycetes), fungi, algae, and viruses.

microirrigation–See **irrigation**, *trickle*.

micronutrient–A plant nutrient found in relatively small amounts(<100 mg kg^{-1}) in plants. These are usually B, Cl, Cu, Fe, Mn, Mo, Ni, Co, and Zn.

microrelief–(i) Generically refers to local, slight irregularities in form and height of a land surface that are superimposed upon a larger landform, including such features as low mounds, swales, and shallow pits. See also **gilgai, shrub-coppice dune, tree-tip mound, tree-tip pit** (ii) Slight variations in the height of a land surface that are too small to delineate on a topographic or soils map at commonly used map scales (e.g. 1:24 000 and 1:15 840).

microsite–A small volume of soil where biological or chemical processes differ from those of the soil as a whole, such as an anaerobic microsite of a soil aggregate or the surface of decaying organic residues.

middlebreaking–See **tillage**, *listing*.

mine dumps–Areas covered with overburden and other waste materials from ore and coal mines, quarries and smelters, and usually with little or no vegetative cover. See also **miscellaneous areas**.

mineral–A naturally occurring homogeneous solid, inorganically formed, with a definite chemical composition and an ordered atomic arrangement.

mineral soil–A soil consisting predominantly of, and having its properties determined predominantly by, mineral matter. Usually contains <200 g kg^{-1} organic carbon (<120-180 g kg^{-1} if saturated with water), but may contain an organic surface layer up to 30 cm thick.

mineralization–The conversion of an element from an organic form to an inorganic state as a result of microbial activity.

mineralogical analysis–The estimation or determination of the kinds or amounts of minerals present in a rock or in a soil.

miner's inch–See **irrigation**, *miner's inch*.

minor elements–See **micronutrients**.

miscellaneous areas–A kind of land area having little or no soil and thus supporting little or no vegetation without major reclamation. Includes areas such as beaches, dumps, rock outcrop, and badlands. The term is used in defining soil survey map units.

miscible displacement–The process that occurs when a fluid mixes with and displaces another fluid. Leaching salts from a soil is an example because the added water mixes with and displaces the soil solution.

mist irrigation–See **irrigation**.

mixed fertilizers–Two or more fertilizer materials mixed or granulated together.

mixing–See **tillage**, *mixing*.

moder–A type of forest humus transitional between mull and mor (term used mostly in Europe; also called **duff mull** in USA and Canada). Sometimes differentiated into the following groups: Mormoder, Leptomoder, Mullmoder, Lignomoder, Hydromoder, and Sapimoder.

moderately-coarse textured–Texture group consisting of coarse sandy loam, sandy loam, and fine sandy loam textures. See also **soil texture**.

moderately-fine textured–Texture group consisting of clay loam, sandy clay loam, and silty clay loam textures. See also **soil texture**.

Mohr circle of stress–A graphical representation of the components of stress acting across the various planes at a given point, drawn with reference to axes of normal stress and shear stress.

Mohr envelope–The envelope of a series of Mohr circles representing stress conditions at failure for a given material.

moisture equivalent–(no longer used in SSSA publications) The weight percentage of water retained by a previously saturated sample of soil 1 cm in thickness after it has been subjected to a centrifugal force of one thousand times gravity for 30 min.

moisture-release curve–(no longer used in SSSA publications) See **soil water**.

moisture-retention curve–See **soil water characteristic**.

moldboard plowing–See **tillage**, *plowing*.

mollic epipedon–A surface horizon of mineral soil that is dark colored and relatively thick, contains at least 5.8 g kg^{-1} organic carbon, is not massive and hard or very hard when dry, has a base saturation of >50% when measured at pH 7, has <110 mg P kg^{-1} soluble in 0.05 M citric acid, and is dominantly saturated with divalent cations.

Mollisols–Mineral soils that have a **mollic epipedon** overlying mineral material with a base saturation of 50% or more when measured at pH 7. Mollisols may have an **argillic**, **natric**, **albic**, **cambic**, **gypsic**, **calcic**, or **petrocalcic horizon**, a **histic epipedon**, or a **duripan**, but not an **oxic** or **spodic horizon**. (An order in the U.S. system of soil taxonomy.)

montmorillonite–$Si_4Al_{1.5}Mg_{0.5}O_{10}(OH)_2Ca_{0.25}$ An aluminum silicate (**smectite**) with a 2:1 layer structure composed of two silica tetrahedral sheets and a shared aluminum and magnesium octahedral sheet. Montmorillonite has a permanent negative charge that attracts interlayer cations that exist in various degrees of hydration thus causing expansion and collapse of the structure (i.e., shrink-swell). The calcium in the formula above is readily exchangeable with other cations. See also **Appendix I**, **Table A3**.

montmorillonite-saponite group–Replaced by **smectite**. See also **phyllosilicate mineral terminology**.

mor–A type of forest **humus** characterized by an accumulation or organic matter on the soil surface in matted Oe(F) horizons, reflecting the dominant mycogenous decomposers. The boundary between the organic horizon and the underlying mineral soil is abrupt. Sometimes differentiated into the following groups: Hemimor, Humimor, Resimor, Lignomor, Hydromor, Fibrimor, and Mesimor.

moraine–An accumulation of **drift**, with an initial topographic expression of its own, built chiefly by the direct action of glacial ice. Examples are **end**, **ground**, lateral, **recessional**, and **terminal moraines**.

mosaic, aerial–An assemblage of overlapping aerial or space photographs or images whose edges have been matched to form a continuous pictorial representation of a portion of the earth's surface.

most probable number–A method for estimating microbial numbers in soil based on dilution to extinction.

mottled zone–A layer that is marked with spots or blotches of different color or shades of color. The pattern of mottling and the size, abundance, and color contrast of the mottles may vary considerably and should be specified in soil description.

mottles–Spots or blotches of different color or shades of color interspersed with the dominant color.

mucigel–The gelatinous material at the surface of roots grown in nonsterile soil. It includes natural and modified plant **exudates** (more specifically mucilages), bacterial cells, and their metabolic products (e.g., capsules and slimes) as well as colloidal mineral and organic matter from the soil.

muck–Organic soil material in which the original plant parts are not recognizable. Contains more mineral matter and is usually darker in color than peat. See also **muck soil, peat, peat soil**, and **sapric material**.

muck soil–An organic soil in which the plant residues have been altered beyond recognition. The sum of the thicknesses of organic layers is usually greater than the sum of the thicknesses of mineral layers.

mucky peat–Organic soil material in which a significant part of the original plant parts are recognizable and a significant part is not. See also peat and muck.

mudflow–A general term for a mass movement landform and a process characterized by a flowing mass of predominantly fine-grained earth material (particles less than 2 mm comprising more than 50 percent of the solid material) possessing a high degree of fluidity during movement. If more than half of the solid fraction consists of material larger than sand size, **debris flow** is preferred.

mulch–See **tillage**, *mulch*.

mulch farming–See **tillage**, *mulch farming*.

mull–A forest **humus** type characterized by intimate incorporation of organic matter into the upper mineral soil (i.e. a well developed A horizon) in contrast to accumulation on the surface. (Sometimes differentiated into the following Groups: Vermimull, Rhizomull, and Hydromull).

multilevel sampling–Collecting remotely sensed data from different types of platforms with ground data from the same geographic area.

multispectral–Generally used for **remote sensing** in two or more spectral bands, such as visible and IR.

Munsell color system–A color designation system that specifies the relative degrees of the three simple variables of color: hue, value, and chroma. For example: 10YR 6/4 is a color (of soil) with a hue = 10YR, value = 6, and chroma = 4. See also **chroma, hue, value, color**.

muscovite–A clear, dioctahedral layer silicate of the **mica** group with Al^{3+} in the octahedral layer and Si and Al in a ratio of 3:1 in the tetrahedral layer. See also **Appendix I, Table A3**.

mutualism–See **symbiosis**.

mycelium–A mass of interwoven filamentous **hyphae**, such as that of the vegetative portion of the thallus of a fungus.

myco–Prefix designating an association or relationship with a fungus (e.g., mycotoxins are toxins produced by a fungus).

mycorrhiza (pl. **mycorrhizae**)–Literally "fungus root". The association, usually symbiotic, of specific fungi with the roots of higher plants. See also **endomycorrhiza** and **ectomycorrhiza**.

N

narrow row planting–See **tillage**, *narrow row planting*.

natric horizon–A mineral soil horizon that satisfied the requirements of an argillic horizon, but that also has prismatic, columnar, or blocky structure and a subhorizon having >15% saturation with exchangeable Na^+.

natural levee–A long, broad low ridge or embankment of sand and coarse silt, built up by a stream on its flood plain and along both sides of its channel. They are wedge-shaped deposits, of the coarsest suspended-load material, that slope gently away from the stream.

neocutan–A **cutan** with a consistent relationship with natural surfaces of soil material. It does not occur immediately at the surfaces. Similar to **hypo-coating**.

net duty of water–See **irrigation**, *net duty of water*.

net primary productivity (NPP)–Net carbon assimilation by plants. NPP = GPP - respiration losses. NPP can be estimated for a given time period as $\Delta B + L + H$, where ΔB = biomass accumulation for the period, L = biomass of material produced in the period and shed (i.e. foliage, flowers, branches), and H = biomass produced in the period and consumed by animals and insects.

neutral soil–A soil in which the surface layer, at least in the tillage zone, is in the pH 6.6 to 7.3 range. See also **acid soil**, **alkaline soil**, **pH**, and **reaction, soil**.

neutralism–A lack of interaction between two organisms in the same ecosystem.

neutron probe–Probe, with radioactive source, that measures soil water content through reflection of scattered neutrons by hydrogen atoms in soil water.

niche–(i) The particular role that a given species plays in the ecosystem; (ii) The physical space occupied by an organism.

nitrate reduction (biological)–The process whereby nitrate is reduced by plants and microorganisms to ammonium for cell synthesis (nitrate assimilation, assimilatory nitrate reduction) or to nitrite by bacteria using nitrate as the terminal electron acceptor in anaerobic respiration (respiratory nitrate reduction, dissimilatory nitrate reduction). Sometimes used synonymously with **"denitrication."**

nitric phosphates–Fertilizers made by processes involving treatment of phosphate rock with nitric acid or a mixture of nitric, sulfuric, or phosphoric acids, usually followed by ammoniation. Water solubility of the phosphorus content may vary over a wide range.

nitrification–Biological oxidation of ammonium to nitrite and nitrate, or a biologically induced increase in the oxidation state of nitrogen.

nitrogen cycle–The sequence of biochemical changes undergone by nitrogen wherein it is used by a living organism, transformed upon the death and decomposition of the organism, and converted ultimately to its original oxidation state.

nitrogen fixation–See **dinitrogen fixation**.

nitrogenase–The specific enzyme system required for biological dinitrogen fixation.

nodule–(i) A cemented concentration of a chemical compound, such as calcium carbonate or iron oxide, that can be removed from the soil intact and that has no orderly internal organization. (ii) [micromorphological] A glaebule with undifferentiated fabric. (iii) Specialized tissue enlargements, or swellings, on the roots, stems, or leaves of plants, such as are caused by nitrogen-fixing microorganisms.

nodule bacteria–The bacteria that fix dinitrogen (N_2) within organized structures (nodules) on the roots, stems, or leaves of plants. Sometimes used as a synonym for "**rhizobia**."

non-inversive tillage–See **tillage**, *non-inversive tillage*.

nonlimiting water range–The region bounded by the upper and lower **soil water** content over which water, oxygen, and mechanical resistance are not limiting to plant growth. Compare with **available water**.

nonpressure solution–Usually nitrogen fertilizer solutions of such low free NH_3 content that no vapor pressure develops and application can be made without need for controlling vapor pressure.

nose slope–The projecting end of an interfluve, where contour lines connecting the opposing side slopes form convex curves around the projecting end and lines perpendicular to the contours diverge downward. Overland flow of water is divergent.

no-till–See **tillage**, *zero tillage*.

nozzle–See **irrigation**, *sprinkler irrigation systems.*

nutrient–Elements or compounds essential as raw materials for organism growth and development.

nutrient antagonism The depressing effect caused by one or more plant nutrients on the uptake and availability of another nutrient in the plant.

nutrient balance–An undefined theoretical ratio of two or more plant nutrient concentrations for an optimum growth rate and yield. Nitrogen and sulfur is an classic example that can be defined because both nutrients are metabolically related in the protein fraction.

nutrient concentration vs. content–Concentration is usually expressed in grams per kilogram (g kg^{-1}) or milligrams per kilogram (mg kg^{-1}) of dry or fresh weight; content is usually expressed as weight per unit area (e.g., kg ha^{-1}). These terms should not be used interchangeably with regard to nutrients in plants.

nutrient deficiency–A low concentration of an essential element that reduces plant growth and prevents completion of the normal plant life cycle.

nutrient efficient plant–A plant that absorbs, translocates, or utilizes more of a specific nutrient than another plant under conditions of relatively low nutrient availability in the soil or growing media.

nutrient interaction–A term usually used to describe the response from two or more nutrients applied together that deviates from additive individual responses when applied separately. This term may also be used to describe metabolic or ion-uptake phenomenon.

nutrient stress–A condition occurring when the quantity of nutrient available reduces growth. It can be from either a deficient or toxic concentration.

nutrient toxicity–Quality, state or degree of harmful effect from an essential nutrient in sufficient concentrations in the plant.

nutrient-supplying power of soils–The capacity of the soil to supply nutrients to growing plants from the **labile**, **exchangeable**, and the moderately **available** forms. See also **fertility, soil**.

n-value–The relationship between the percentage of water under field conditions and the percentages of inorganic clay and of humus.

O

O horizon–See **soil horizon** and **Appendix II**.

Oa horizon (H layer)–A layer occurring in **mor** humus consisting of well-decomposed organic matter of unrecognizable origin (**sapric material**). See also **soil horizon** and **Appendix II**.

Ochrepts–**Inceptisols** formed in cold or temperate climates and that commonly have an **ochric epipedon** and a **cambic horizon**. They may have an **umbric** or **mollic epipedon** <25 cm thick or a **fragipan** or **duripan** under certain conditions. These soils are not dominated by **amorphous materials** and are not saturated with water for periods long enough to limit their use for most crops. (A suborder in the U.S. system of soil taxonomy.)

ochric epipedon–A surface horizon of mineral soil that is too light in color, too high in chroma, too low in organic carbon, or too thin to be a **plaggen, mollic, umbric, anthropic** or **histic epipedon**, or that is both hard and massive when dry.

Oe horizon (F layer)–A layer of partially decomposed litter with portions of plant structures still recognizable (**hemic material**). Occurs below the L layer on the forest floor in forest soils. It is the fermentation layer. See also **soil horizon** and **Appendix II**.

Oi horizon [L layer (litter)]–A layer of organic material having undergone little or no decomposition (**fibric material**). On the forest floor this layer consists of freshly fallen leaves, needles, twigs, stems, bark, and fruits. This layer may be very thin or absent during the growing season. See also **soil horizon** and **Appendix II**.

oil wasteland–Areas on which liquid oily wastes, principally saltwater and oil, have accumulated. Includes slush pits and adjacent areas affected by oil waste. A **miscellaneous area**.

oligotrophic–Environments in which the concentration of nutrients available for growth is limited. Nutrient poor habitats.

oligotrophs–Organisms able to grow in environments with low nutrient concentrations.

one-third-atmosphere percentage–(no longer used in SSSA publications) The percentage of water contained in a soil that has been saturated, subjected to, and is in equilibrium with, an applied pressure of one-third atmosphere. Approximately the same as **one-third-bar percentage**. See **soil water**, *soil water potential*.

one-third-bar percentage–(no longer used in SSSA publications) The percentage of water contained in a soil that has been saturated, subjected to, and is in equilibrium with, an applied pressure of one-third bar. Approximately the same as **one-third-atmosphere percentage**. See also **soil water**, *soil water potential*.

organan–A **cutan** composed of a concentration of organic matter.

organic farming–Crop production system that reduces, avoids or largely excludes the use of synthetically compound fertilizers, pesticides, growth regulators, and livestock feed additives.

organic fertilizer–By product from the processing of animals or vegetable substances that contain sufficient plant nutrients to be of value as fertilizers.

organic soil–A soil in which the sum of the thicknesses of layers containing organic soil materials is generally greater than the sum of the thicknesses of mineral layers.

organic soil materials–Soil materials that are saturated with water and have 174 g kg^{-1} or more organic carbon if the mineral fraction has 500 g kg^{-1} or more clay, or 116 g kg^{-1} organic carbon if the mineral fraction has no clay, or has proportional intermediate contents, or if never saturated with water, have 203 g kg^{-1} or more organic carbon.

organotroph–See **heterotroph**.

Orthents–**Entisols** that have either textures of very fine sand or finer in the fine earth fraction, or textures of loamy fine sand or coarser and a coarse fragment content of 35% or more and that have an organic carbon content that decreases regularly with depth. **Orthents** are not saturated with water for periods long enough to limit their use for most crops. (A suborder in the U.S. system of soil taxonomy.)

Orthids–Previous to 1994 this term was used to indicate **Aridisols** that have a cambic, calcic, petrocalcic, gypsic, or salic horizon or a duripan but that lack an argillic or natric horizon. The term was dropped as a suborder in the 1994 revision of the USDA, *Soil taxonomy*.

Orthods–**Spodosols** that have less than six times as much free iron (elemental) than organic carbon in the spodic horizon but the ratio of iron to carbon is 0.2 or more. **Orthods** are not saturated with water for periods long enough to limit their use for most crops. (A suborder in the U.S. system of soil taxonomy.)

orthophosphate–A salt of orthophosphoric acid such as $(NH_4)_2HPO_4$, $CaHPO_4$, or K_2HPO_4.

Orthox–**Oxisols** that are moist all or most of the time, and that have a low to moderate content of organic carbon within the upper 1 m or a mean annual soil temperature of 22° C or more. (A suborder in the U.S. system of soil taxonomy.)

ortstein–A cemented **spodic horizon**.

osmotic potential, pressure–See **soil water**.

outer sphere adsorption–Adsorption of ions that occurs with the retention of waters of hydration between the surface and the adsorbed ion where the force that retains the ion is only electrostatic attraction. Ions that are retained by outer sphere adsorption are readily exchangeable. See also **exchangeable cation** and **exchangeable anion**.

outwash–Stratified detritus (chiefly sand and gravel) removed or "washed out" from a **glacier** by melt-water streams and deposited in front of or beyond the **end moraine** or the margin of an active glacier. The coarser material is deposited nearer to the ice.

oven-dry soil–Soil that has been dried at 105° C until it reaches constant mass.

overburden–(i) Recently transported and deposited material that occurs immediately superjacent to the surface horizon of a contemporaneous soil. (ii) A term used to designate disturbed or undisturbed material of any nature, consolidated or unconsolidated, that overlies a deposit of useful materials, ores, lignites, or coals, especially those deposits mined from the surface by open cuts.

overconsolidated soil deposit–A soil deposit that has been subjected to an effective pressure greater than the present over-burden pressure.

overlay–(i) A transparent sheet giving information to supplement that shown on maps. When the overlay is laid over the map on which it is based, its details will supplement the map. (ii) A tracing of selected details on a photograph, **mosaic**, or map to present the interpreted features and the pertinent detail, or to facilitate plotting.

oxbow lake–The crescent-shaped, often **ephemeral** body of standing water situated by the side of a stream in the abandoned channel (oxbow) of a meander after the stream formed a neck cutoff and the ends of the original bend were silted up.

oxic horizon–A mineral soil horizon that is at least 30 cm thick and characterized by the virtual absence of weatherable primary minerals or 2:1 layer silicate clays, the presence of 1:1 layer silicate clays and highly insoluble minerals such as quartz sand, the presence of hydrated oxides of iron and aluminum, the absence of water-dispersible clay, and the presence of low cation exchange capacity and small amounts of exchangeable bases.

oxidation–The loss of one or more electrons by an ion or molecule.

oxidation ditch–An artificial open channel for partial digestion of liquid organic wastes in which the wastes are circulated and aerated by a mechanical device.

oxidation state–The number of electrons to be added (or subtracted) from an atom in a combined state to convert it to the elemental form.

oxidation-reduction potential–See E_H and **pe**.

oxidative phosphorylation–Conversion of inorganic phosphate into the energy-rich phosphate of adenosine 5′-triphosphate.

Oxisols–Mineral soils that have an **oxic horizon** within 2 m of the surface or plinthite as a continuous phase within 30 cm of the surface, and that do not have a **spodic** or **argillic horizon** above the **oxic horizon**. (An order in the U.S. system of soil taxonomy.)

oxyaquic conditions–Pertaining to soils that are saturated but are not reduced and do not contain redoximorphic features.

oxytropic–The response of a biological organism to the presence of oxygen.

P

P_2O_5–Phosphorus pentoxide; designation on the fertilizer label that denotes the percentage of available phosphorus reported as phosphorus pentoxide.

packing voids (compound)–Voids formed by the random packing of peds that do not accommodate each other.

packing voids (simple)–Voids formed by the random packing of single skeletal grains.

paleosol–A soil that formed on a landscape in the past with distinctive morphological features resulting from a soil-forming environment that no longer exists at the site. The former pedogenic process was either altered because of external environmental change or interrupted by burial. A paleosol (or component horizon) may be classed as relict if it has persisted in a land-surface position without major alteration of morphology by processes of the prevailing pedogenic environment. An exhumed paleosol is one that formerly was buried and has been re-exposed by erosion of the covering mantle.

palygorskite–$Si_8Mg_2Al_2O_{20}(OH)_2(OH_2)$. $4H_2O$ (i) A fibrous clay mineral composed of two silica tetrahedral sheets and one aluminum and magnesium octahedral sheet that make up the 2:1 layer that occurs in strips. The strips that have an average width of two linked tetrahedral chains are linked at the edges forming tunnels where water molecules are held. Palygorskite is most common in soils of arid regions. Also referred to as **attapulgite**. (ii) A magnesium aluminum silicate clay used in fertilizer production, including conditioning of fertilizer products, and as a suspending agent in suspension fertilizers.

pan–(i) Genetic - A natural subsurface soil layer with low or very low **hydraulic conductivity** and differing in certain physical and chemical properties from the soil immediately above or below the pan. See **caliche, claypan, fragipan**, and **hardpan**, all of which are genetic pans.(ii) Pressure or Induced - A subsurface horizon or soil layer having a higher **bulk density** and a lower total **porosity** than the soil directly above or below it, as a result of pressure that has been applied by normal tillage operations or by other artificial means. Frequently referred to as **plowpan**, plowsole, or **traffic pan**. See also **tillage**, *plow pan* and **tillage**, *pressure pan.*

papule–Glaebule composed dominantly of clay minerals with continuous and/or lameliar fabric, and sharp external boundaries.

paralithic contact–Similar to a lithic contact except that it is softer, can be dug with difficulty with a spade, if a single mineral has a hardness <3 (Mohs scale), and gravel size chunks will partially disperse within 15 hours shaking in water or sodium hexametaphosphate solution.

paraplow–See **tillage**, *paraplow.*

parasitism–(i) Feeding by one organism on the cells of a second organism, which is usually larger than the first. The parasite is, to some extent, dependent on the host at whose expense it is maintained. (ii) An association whereby one organism (parasite) lives in or on another organism (host) and benefits at the expense of the host.

paratill–See **tillage**, *paratill.*

parent material–The unconsolidated and more or less chemically weathered mineral or organic matter from which the **solum** of soils is developed by pedogenic processes.

parna–A term used, especially in southeast Australia, for silt and sand-sized aggregates of **eolian** clay occurring in sheets.

particle density–The density of the soil particles, the dry mass of the particles being divided by the solid (not bulk) volume of the particles, in contrast with bulk density. Units are Mg m^{-3}.

particle size–The effective diameter of a particle measured by sedimentation, sieving, or micrometric methods.

particle size analysis–Determination of the various amounts of the different **soil separates** in a soil sample, usually by sedimentation, sieving, micrometry, or combinations of these methods.

particle size distribution–The fractions of the various **soil separates** in a soil sample, often expressed as mass percentages.

parts per million (ppm)–(no longer used in SSSA publications) (i) The concentration of solutions expressed in weight or mass units of solute (dissolved substance) per million weight or mass units of solution. (ii) A concentration in solids expressed in weight or mass units of a substance contained per million weight or mass units of solid, such as soil.

pasteurization–Partial sterilization of soil, liquid, or other natural substances by temporary heat treatment.

patterned ground–A general term for any ground surface exhibiting a discernibly ordered, more-or-less symmetrical, morphological pattern of ground and, where present, vegetation. Patterned ground is characteristic of, but not confined to, permafrost regions or areas subjected to intense frost action; it also occurs in tropical, subtropical, and temperate areas. Patterned ground is classified by type of pattern and presence or absence of sorting and includes nonsorted and sorted circles, net, polygons, steps and stripes, garlands, and **solifluction** features. In **permafrost** regions, the most common macroform is the ice-wedge polygon and a common microform is the nonsorted circle.

PCR (polymerase chain reaction)–An *in vitro* method for amplifying defined segments of DNA. PCR involves a repeated cycle of oligonucleotide **hybridization** and extension on single-stranded DNA templates.

pe–The negative logarithm of the apparent electron activity which can be calculated by including the apparent activity of electrons in equilibrium calculations of **redox** half-cell reactions. In practice it is used as an alternative to E_H and at $25°$ C can be calculated from E_H values expressed in volts by dividing by 0.059.

peat–Organic soil material in which the original plant parts are recognizable (**fibric material**). See also **peat soil**, **muck**, **muck soil**, and **Histosol**.

peatland–A generic term for any wetland that accumulates partially decayed plant matter. Mire, moor and muskeg are terms for European and Canadian peatlands. See also **bog** and **fen**.

peat soil–An organic soil in which the plant residues are recognizable. The sum of the thicknesses of the organic layers are usually greater than the sum of the thicknesses of the mineral layers. See also **peat**, **muck**, **muck soil**, and **Histosol**.

pebbles–Rounded or partially rounded rock or mineral fragments between 2 and 75 mm in diameter. Size may be further refined as fine pebbles (2-5 mm diameter), medium pebbles (5-20 mm diameter), and coarse pebbles (20-75 mm diameter). See also **rock fragments**.

ped–A unit of soil structure such as a block, column, granule, plate, or prism, formed by natural processes (in contrast with a **clod**, which is formed artificially). See also **shrinkage, soil**, *ped (shrinkage)*.

pedal–Applied to soil materials, most of which consists of **peds**.

pedalfer–A subdivision of a soil order comprising a large group of soils in which sesquioxides increased relative to silica during soil formation. (Not used in current U.S. system of soil taxonomy.)

pediment–A gently sloping, erosional surface developed at the foot of a receding hill or mountain slope. The surface can be bare or it may be thinly mantled with **alluvium** and **colluvium** in transport to the adjacent valley.

pediplain–A geomorphic term for an **outwash** plain landform.

pedisediment–A layer of sediment, eroded from the **shoulder** and **back slope** of an erosional slope, that lies on and is, or was, being transported across a **pediment**.

pedocal–A subdivision of a soil order comprising a large group of soils in which calcium accumulated during soil formation. (Not used in current U.S. system of soil taxonomy.)

pedological features–Recognizable units within a soil material which are distinguishable from the enclosing material for any reason such as origin (deposition as an entity), differences in concentration of some fraction of the plasma, or differences in arrangement of the constituents (fabric).

pedon–A three-dimensional body of soil with lateral dimensions large enough to permit the study of horizon shapes and relations. Its area ranges from 1 to 10 m^2. Where horizons are intermittent or cyclic, and recur at linear intervals of 2 to 7 m, the pedon includes one-half of the cycle. Where the cycle is <2 m, or all horizons are continuous and of uniform thickness, the pedon has an area of approximately 1 m^2. If the horizons are cyclic, but recur at intervals >7 m, the pedon reverts to the 1 m^2 size, and more than one soil will usually be represented in each cycle.

pedotubule–A pedological feature consisting of soil material and having a tubular external form and relatively sharp external boundaries.

peneplain–A once high, rugged area which has been reduced by erosion to a low, gently rolling surface resembling a plain.

penetrability–The ease with which a probe can be pushed into the soil. (May be expressed in units of distance, speed, force, or work depending on the type of penetrometer used.)

penetration resistance–The force per unit area on a standard ASAE cone necessary for penetration by the cone. See also **cone index**.

penetrometer–See **cone penetrometer**.

percent area wetted–See *irrigation, percent area wetted.*

percolation, soil water–The downward movement of water through soil. Especially, the downward flow of water in saturated or nearly saturated soil at **hydraulic gradients** of the order of 1.0 or less.

pergelic–A soil temperature regime that has mean annual soil temperatures of <0°C. **Permafrost** is present. See also **permafrost**.

periglacial–Pertaining to processes, conditions, areas, climates, and topographic features occurring at the immediate margins of **glaciers** and ice sheets, and influenced by cold temperature of the ice.

permafrost–(i) Permanently frozen material underlying the solum. (ii) A perennially frozen soil horizon.

permafrost table–The upper boundary of the permafrost coincident with the lower limit of seasonal thaw. See also **permafrost** (i).

permanent charge–The net negative (or positive) charge of clay particles inherent in the crystal structure of the particle; not affected by changes in pH or by ion-exchange reactions.

permanent wilting point–The largest water content of a soil at which indicator plants, growing in that soil, wilt and fail to recover when placed in a humid chamber. Often estimated by the water content at -1.5 MPa soil matric potential.

permeability, soil–(i) The ease with which gases, liquids, or plant roots penetrate or pass through a bulk mass of soil or a layer of soil. Since different soil horizons vary in permeability, the particular horizon under question should be designated. (ii) The property of a porous medium itself that expresses the ease with which gases, liquids, or other substances can flow through it, and is the same as intrinsic permeability k. See also **intrinsic permeability**, **Darcy's law**, and **soil water**.

permeameter–A device for confining a sample of soil or porous medium and subjecting it to fluid flow, in order to measure the **hydraulic conductivity** or **intrinsic permeability** of the soil or porous medium for the fluid.

Perox–**Oxisols** that have a **perudic** soil moisture regime. (A suborder in the U.S. system of soil taxonomy.)

perudic–A **udic** soil moisture regime in which water moves through the soil in all months when it is not frozen. (A suborder in the U.S. system of soil taxonomy.)

petrocalcic horizon–A continuous, indurated calcic horizon that is cemented by calcium carbonate and, in some places, with magnesium carbonate. It cannot be penetrated with a spade or auger when dry, dry fragments do not slake in water, and it is impenetrable to roots.

petroferric contact–A boundary between soil and a continuous layer of indurated soil in which iron is an important cement. Contains little or no organic matter.

petrogypsic horizon–A continuous, strongly cemented, massive, gypsic horizon that is cemented by calcium sulfate. It can be chipped with a spade when dry. Dry fragments do not slake in water and it is impenetrable to roots.

pH, soil–The pH of a solution in equilibrium with soil. It is determined by means of a glass, quinhydrone, or other suitable electrode or indicator at a specified soil-solution ratio in a specified solution, usually distilled water, 0.01 M $CaCl_2$, or 1 M KCl.

pH$_c^*$ –The calculated pH that a solution would have if it were in equilibrium with calcium carbonate. Numerically, pH$_c^*$ is equal to $(pK_2 - pK_c) + p(Ca) + pAlk$, where p(Ca) and pAlk are the negative logarithms of the molar concentrations of Ca and of the equivalent concentration of $(CO_3 + HCO_3)$, respectively, and pK_2 and pK_c are the negative logarithms of the second dissociation constant of H_2CO_3 and the solubility constant of $CaCO_3$, respectively, both corrected for ionic strength. It is used in conjunction with the measured pH of a water to determine if $CaCO_3$ will precipitate from the water, or if the water will dissolve $CaCO_3$ as it passes through a calcareous soil.

pH-dependent charge–The portion of the **cation** or **anion exchange capacity** which varies with pH. See also **acidity, residual** and **variable charge**.

phase–A utilitarian grouping of soils defined by soil or environmental features that are not class differentia used in U.S. system of soil taxonomy, e.g., surface texture, surficial rock fragments, rock outcrops, substratum, special soil water conditions, salinity, physiographic position, erosion, thickness, etc. Phase identifications are introduced into soil names by adding them to a taxon name as modifiers. See also **taxon, component soil**.

phosphate–In fertilizer trade terminology, phosphate is used to express the sum of the water-soluble and the citrate-soluble phosphoric acid (P_2O_5); also referred to as the available phosphoric acid (P_2O_5).

phosphate rock–A microcrystalline, calcium fluorophosphate of sedimentary or igneous origin of varying P content. It is usually concentrated and solubilized to be used directly or concentrated in manufacture of commercial phosphate fertilizers.

phosphobacteria–Bacteria able to convert organic phosphorus into orthophosphate.

phosphoric acid–In commercial fertilizer manufacturing, it is used to designate orthosphorphoric acid, H_3PO_4. In fertilizer labeling, it is the common term used to represent the phosphate concentration in terms of available P, expressed as percent P_2O_5.

phosphorus, fixation–(no longer used in SSSA publications) The immobilization of phosphorus by strong adsorption or precipitation.

photolithotroph–An organism that uses light as a source of energy and CO_2 or carbonates as the source of carbon for cell biosynthesis. See also **autotroph**.

photomap–A **mosaic map** made from aerial photographs with physical and cultural features, marginal data, and other map information as shown on a planimetric map.

phototropic–The response of a biological organism to the presence of light.

phyllosilicate mineral terminology–Phyllosilicate minerals have layer structures composed of shared octahedral and tetrahedral sheets. See also **Appendix I, Table A3**.

plane (of atoms)–A flat (planar) array of atoms of one atomic thickness. Example: plane of basal oxygen atoms within a tetrahedral sheet.

sheet (of polyhedra)–Flat array of more than one atomic thickness and composed of one level of linked coordination polyhedra. A sheet is thicker than a plane and thinner than a layer. Example: tetrahedral sheet, octahedral sheet.

layer–A combination of sheets in a 1:1 or 2:1 assemblage.

interlayer–Materials between structural layers of minerals, including cations, hydrated cations, organic molecules, and hydroxide octahedral groups and sheets.

unit structure–The total assembly of a layer plus interlayer material.

phyllosphere–The surface of above-ground living plant parts.

physical properties (of soils)–Those characteristics, processes, or reactions of a soil which are caused by physical forces and which can be described by, or expressed in, physical terms or equations. Examples of physical properties are **bulk density, hydraulic conductivity, porosity, pore-size distribution**, etc.

physical weathering–The breakdown of rock and mineral particles into smaller particles by physical forces such as frost action. See also **weathering**.

physiosorption–(no longer used in SSSA publications) The process of attachment of non-ionic substances such as polar water molecules, acetic acid molecules, or nucleic acids to clays or to other solid-phase surfaces. The attachment of large molecules to clay particles by ionic processes is *not* physiosorption.

phytoliths–Inorganic bodies derived from replacement of plant cells; they are usually opaline.

phytometer–A plant or plants used to measure the physical factors of the habitat in terms of physiological activities.

phytomorphic soils–(Canada) Well-drained soils of an association which have developed under the dominant influence of the natural vegetation characteristic of a region. The **zonal soils** of an area.

phytotoxic–The property of a substance at a specified concentration that restricts or constrains plant growth.

pit and mound topography–Complex microrelief created by numerous **cradle knolls** and their attendant pits. Usually associated with forested sites or cleared sites that have not been plowed. See also **microrelief**.

pits–Open excavations from which soil and commonly, underlying material, have been removed exposing either rock or other material that supports few or no plants. Includes mine pits, gravel pits, and quarry pits. A **miscellaneous area**.

placic horizon–A black to dark reddish mineral soil horizon that is usually thin but that may range from 1 mm to 25 mm in thickness. The placic horizon is commonly cemented with iron and is slowly permeable or impenetrable to water and roots.

plaggen epipedon–A man-made surface horizon more than 50 cm thick that is formed by long-continued manuring and mixing.

Plaggepts–**Inceptisols** that have a **plaggen epipedon**. (A suborder in the U.S. system of soil taxonomy.)

plain–A flat, undulating, or even rolling area, larger or smaller, that includes few prominent hills or valleys, that usually is at low elevation in reference to surrounding areas, and that may have considerable overall slope and local relief.

plane (of atoms)–See **phyllosilicate mineral terminology**.

Planosol–A great soil group of the **intrazonal** order and **hydromorphic** suborder consisting of soils with **eluviated** surface horizons underlain by B horizons more strongly eluviated, cemented, or compacted than associated normal soil. (Not used in current U.S. system of soil taxonomy.)

plant analysis–The determination of the nutrient concentration in plants or plant parts with analytical procedures.

plant food–The inorganic compounds elaborated within a plant to nourish its cells; a frequent synonym for plant nutrients, particularly in the fertilizer trade.

plant nutrient–An element which is absorbed by plants and is necessary for completion of the normal life cycle. These include C, H, O, N, P, K, Ca, Mg, S, Cu, Fe, Zn, Mn, B, Cl, Ni, and Mo.

plasma–That part of the soil material that is capable of being or has been moved, reorganized, and/or concentrated by the processes of soil formation. It includes all the material, mineral or organic, of colloidal size and relatively soluble material that is not contained in the skeleton grains.

plasmic fabric–The arrangement of **plasma, skeleton grains**, and associated simple **packing voids**.

plasmids–Extrachromosomal DNA.

plasticity constants–See **Atterberg limits, consistency, liquid limit, plastic limit**, and **plasticity number**.

plastic limit–The minimum water mass content at which a small sample of soil material can be deformed without rupture. Synonymous with "lower plastic limit." See also **Atterberg limits, consistency, liquid limit**, and **plasticity number**.

plastic soil–A soil capable of being molded or deformed continuously and permanently, by relatively moderate pressure, into various shapes. See also **consistence**.

plasticity number–The numerical difference between the liquid and the plastic limit or, synonymously, between the lower plastic limit and the upper plastic limit. Sometimes called "plasticity index." See also **Atterberg limits**, **consistency**, **liquid limit**, and **plastic limit**.

plasticity range–The range of water mass content within which a small sample of soil exhibits plastic properties.

plate count–A count of the number of colonies formed on a solid culture medium when inoculated with a small amount of soil. The technique has been used to estimate the number of certain organisms present in the soil sample.

platy soil structure–A shape of soil structure. See also **soil structure** and **soil structure shapes**.

platy–Consisting of soil aggregates that are developed predominantly along the horizontal axes; laminated; flaky. See also **soil structure types**.

playa–An ephemerally flooded, vegetatively barren area on a basin floor that is veneered with fine-textured sediment and acts as a temporary or as the final sink for drainage water. See also **miscellaneous areas**.

plinthite–A weakly-cemented iron-rich, humus poor mixture of clay with other diluents that commonly occurs as dark red redox concentrations that form platy, polygonal, or reticulate patterns. Plinthite changes irreversibly to ironstone hardpans or irregular aggregates on exposure to repeated wetting and drying.

plow layer–See **tillage**, *plow layer*.

plow pan–See **tillage**, *plow pan*.

plowing–See **tillage**, *plowing*.

plowless farming–See **tillage**, *plowless farming*.

plow-planting–See **tillage**, *plow-planting*.

pocosin–A swamp, usually containing organic soil, and partly or completely enclosed by a sandy rim. The Carolina Bays of the Southeastern USA.

Podzol–A great soil group of the **zonal** order consisting of soils formed in cool-temperate to temperate, humid climates, under coniferous or mixed coniferous and deciduous forest, and characterized particularly by a highly-leached, whitish-gray (**Podzol**) A2 (E) horizon. (Not used in current U.S. system of soil taxonomy.)

podzolization–A process of soil formation resulting in the genesis of **Podzols** and Podzolic soils.

point bar–One of a series of low, arcuate ridges of sand and gravel developed on the inside of a growing meander by the slow addition of individual accretions accompanying migration of the channel toward the outer bank.

point of zero net charge (pznc)–The pH value of a solution in equilibrium with a variable charge material or mixture of materials whose net charge from all sources is zero (i.e. anion exchange capacity = effective cation exchange capacity). It is often determined for soils that are low in permanent charge minerals and high in oxides and hydrous oxides of Fe and Al.

pollution–The presence or introduction of a pollutant into the environment.

polypedon–A group of contiguous similar pedons. The limits of a polypedon are reached at a place where there is no soil or where the pedons have characteristics that differ significantly.

pore ice–Frozen water in the interstitial pores of a porous medium.

pore-size distribution–The volume fractions of the various size ranges of pores in a soil, expressed as percentages of the soil bulk volume (soil particles plus pores). See also **Table 2**.

pore space–The portion of soil bulk volume occupied by soil pores.

pore volume–See **pore space**.

pore water velocity–The velocity at which water travels in pores relative to a given axis. It is equal to the **flux density** divided by the **soil water** content.

porosity–The volume of pores in a soil sample (nonsolid volume) divided by the **bulk volume** of the sample.

porous trickle tubing–See **irrigation,** *trickle.*

potash–Term used to refer to potassium or potassium fertilizers and usually designated as K_2O.

potassium fixation–The process of converting exchangeable or water-soluble potassium to that occupying the position of K^+ in the micas. They are counter-ions entrapped in the ditrigonal voids in the plane of basal oxygen atoms of some phyllosilicates as a result of contraction of the interlayer space. The fixation may occur spontaneously with some minerals in aqueous suspensions or as a result of heating to remove interlayer water in others. Fixed K^+ ions are exchangeable only after expansion of the interlayer space. See also **ammonium fixation**.

pothole–Shallow marsh-like ponds, particularly as found in the Dakotas.

Prairie soils–A **zonal** great soil group consisting of soils formed under temperate to cool-temperate, humid regions under tall grass vegetation. (Not used in current U.S. system of soil taxonomy.)

precipitation interception–The stopping, interrupting, or temporary holding of descending precipitation in any form by mulch, a vegetative canopy, vegetation residue or any other physical barrier.

predation–A relationship between two organisms whereby one organism (predator) engulfs and digests the second organism (prey).

preferential flow–The process whereby free water and its constituents move by preferred pathways through a porous medium. Also called **bypass flow**.

preplant irrigation–See **irrigation,** *preplant irrigation.*

pressure membrane–A membrane, permeable to water and only very slightly permeable to gas when wet, through which water can escape from a soil sample in response to a pressure gradient.

primary mineral–A mineral that has not been altered chemically since deposition and crystallization from molten lava. See also **secondary mineral**.

primary nutrients–Refers to N, P, and K in fertilizers. See also **macronutrient**.

priming effect–Stimulation of microbial activity in soil, usually organic matter decomposition, by the addition of labile organic matter.

prismatic soil structure–A shape of soil structure. See also **soil structure** and **soil structure shapes**.

procaryotes–See **prokaryotes**.

productive capacity–See **soil productivity**.

Table 2. Pore size classification. (after Brewer, R. 1964. Fabric and mineral analysis of soils, John Wiley & Sons)

Class	Subclass	Class limits equivalent diameter (μm)
Macropores	Coarse	>5000
	Medium	2000-5000
	Fine	1000-2000
	Very Fine	75-1000
Mesopores		30-75
Micropores		5-30
Ultramicropores		0.1-5
Cryptopores		<0.1

productivity, soil–The output of a specified plant or group of plants under a defined set of management practices.

profile, soil–A vertical section of the soil through all its horizons and extending into the C horizon.

prokaryotes (procaryotes)–A cell or organism lacking a true nucleus.

propagule–Any cell unit capable of developing into a complete organism. For fungi, the unit may be a single spore, a cluster of spores, hyphae, or a hyphal fragment.

protocooperation–An association of mutual benefit to two or more species but without the cooperation being obligatory for their existence or the performance of some function.

proximal–Said of a sedimentary deposit consisting of coarse **clastics** and deposited nearest the source area. See also **distal**.

Psamments–**Entisols** that have textures of loamy fine sand or coarser in all parts, have <35% coarse fragments, and that are not saturated with water for periods long enough to limit their use for most crops. (A suborder in the U.S. system of soil taxonomy.)

psychrophile–See **psychrophilic organism.**

psychrophilic organism–An organism whose optimum temperature for growth falls in the approximate range of 5 to 15° C. Synonymous with **cryophile**.

pure culture–A population of microorganisms composed of a single strain. Such cultures are obtained through selective laboratory procedures and are rarely found in a natural environment.

pyroclastics–A general term applied to detrital volcanic materials that have been explosively or aerially ejected from a volcanic vent.

pyrophosphate–A class of phosphorus compounds produced by the reaction of either anhydrous ammonia or potassium hydroxide with pyrophosphoric acid ($H_4P_2O_7$). Pyrophosphoric acid is a condensation product of two molecules of orthophosphoric acid (H_3PO_4). The main polyphosphate species in polyphosphate fertilizers.

pyrophyllite–$Si_4Al_2O_{10}(OH)_2$ An aluminosilicate mineral with a 2:1 layer structure but without isomorphous substitution. It is dioctahedral. See also **Appendix I, Table A3**.

Q

quantity intensity ratio–The change in quantity sorbed with change in quantity in solution. It is determined from the slope of the plot of concentration in solution vs. the quantity sorbed. See **sorption**.

R

R layer–See **soil horizon** and **Appendix II**.

rainfall erosivity index–See **erosion**, *rainfall erosivity index*.

rainfall interception–See **precipitation interception**.

raised bed–See **bed**.

reaction, soil–(no longer used in SSSA publications) The degree of acidity or alkalinity of a soil, usually expressed as a pH value. Descriptive terms commonly associated with certain ranges in pH are: *extremely acid*, <4.5; *very strongly acid*, 4.5-5.0; *strongly acid*, 5.1-5.5; *moderately acid*, 5.6-6.0; *slightly acid*, 6.1-6.5; *neutral*, 6.6-7.3; *slightly alkaline*, 7.4-7.8; *moderately alkaline*, 7.9-8.4; *strongly alkaline*, 8.5-9.0; and *very strongly alkaline*, >9.1.

recessional moraine–An **end** or lateral **moraine**, built during a temporary but significant halt in the retreat of a **glacier**. Also, a moraine built during a minor readvance of the ice front during a period of recession. See also **end moraine, ground moraine, terminal moraine**.

Red Desert soil–A **zonal** great soil group consisting of soils formed under warm-temperate to hot, dry regions under desert-type vegetation, mostly shrubs. (Not used in current U.S. system of soil taxonomy.)

red earth–Highly leached, red clayey soils of the humid tropics, usually with very deep profiles that are low in silica and high in **sesquioxides**. (Not used in current U.S. system of soil taxonomy.)

redistribution (of soil water)–The process of soil-water movement to achieve an equilibrium energy state of water throughout the soil.

redox–Reduction-oxidation.

redox concentrations–Zones of apparent accumulation of Fe-Mn oxides in soils.

redox depletions–Zones of low chroma (2 or less) where Fe-Mn oxides alone or both Fe-Mn oxides and clay have been stripped out of the soil.

redoximorphic features–Soil properties associated with wetness that result from the reduction and oxidation of iron and manganese compounds in the soil after saturation with water and desaturation, respectively. See also **redox concentrations, redox depletions**.

redox-potential–See E_H and **pe**.

reduced matrix–A soil matrix which has a low chroma in situ, but undergoes a change in hue or chroma within 30 minutes after the soil material is exposed to air. The color change is due to the oxidation of iron.

reduction–The gain of one or more electrons by an ion or molecule.

Red-Yellow Podzolic soils–A combination of the **zonal** great soil groups, Red Podzolic and Yellow Podzolic, consisting of soils formed under warm-temperate to tropical, humid climates, under deciduous or coniferous forest vegetation and usually, except for a few members of the Yellow Podzolic Group, under conditions of good drainage. (Not used in current U.S. system of soil taxonomy.)

reel and gun irrigation (traveling gun)–See **irrigation.**

reference electrode–An electrode that maintains an invariant potential under the conditions prevailing in an electrochemical measurement and thereby permits measurement of the potential of an **ion-selective** or a platinum (redox) electrode.

reflectance–The ratio of the radiant energy reflected by a body to that incident upon it. The suffix (-ance) implies a property of that particular specimen surface.

regolith–The unconsolidated mantle of weathered rock and soil material on the earth's surface; loose earth materials above solid rock. (Approximately equivalent to the term "soil" as used by many engineers.)

Regosol–Any soil of the **azonal** order without definite genetic horizons and developing from or on deep, unconsolidated, soft mineral deposits such as sands, loess, or glacial drift. (Not used in current U.S. system of soil taxonomy.)

Regur–An **intrazonal** group of dark calcareous soils high in clay, which is mainly montmorillonitic, and formed mainly from rocks low in quartz; occurring extensively on the Deccan Plateau of India. (Not used in current U.S. system of soil taxonomy.)

relative yield–The harvestable or biomass yield with or without supplementation of the nutrient in question expressed as a percentage of the yield with the nutrient in adequate amounts.

relief–The relative difference in elevation between the upland summits and the lowlands or valleys of a given region.

remote sensing–Refers to the full range of activities that collects information from a distance, e.g., the utilization at a distance (as from aircraft, spacecraft, or ship) of any device for measuring electromagnetic radiation, force fields, or acoustic energy. The technique employs such devices as the camera, lasers, and radio frequency receivers, radar systems, sonar, seismographs, gravimeters, magnetometers, and scintillation counters.

Rendolls–**Mollisols** that have no **argillic** or **calcic horizon** but that contain material with $CaCO_3$ equivalent >400 g kg^{-1} within or immediately below the **mollic epipedon. Rendolls** are not saturated with water for periods long enough to limit their use for most crops. (A suborder in the U.S. system of soil taxonomy.)

Rendzina–A great soil group of the **intrazonal** order and calcimorphic suborder consisting of soils with brown or black friable surface horizons underlain by light gray to pale yellow calcareous material developed from soft, highly calcareous parent material under grass vegetation or mixed grasses and forest in humid and semiarid climates. (Not used in current U.S. system of soil taxonomy.)

reservoir tillage–See **tillage.**

residual fertility–The available nutrient content of a soil carried over to subsequent crops.

residual material–Unconsolidated and partly weathered mineral materials accumulated by disintegration of consolidated rock in place.

residual shrinkage–See **shrinkage, soil.**

residual soil–A soil formed from, or resting on, consolidated rock of the same kind as that from which it was formed, and in the same location. (Not used in current U.S. system of soil taxonomy.)

residue processing–See **tillage**, *residue processing*.

residuum–Unconsolidated, weathered, or partly weathered mineral material that accumulates by disintegration of bedrock in place. See also **saprolite, regolith, colluvium**.

resolution–The ability of an entire remote sensor system, including lens, antennae, display, exposure, processing, and other factors, to render a sharply defined image.

respiratory quotient (RQ)–The number of molecules of CO_2 liberated for each molecule of O_2 consumed.

restriction enzyme–A class of highly specific enzymes which make double stranded breaks in DNA at specific sites near where they combine.

retardation factor–The capability of a soil for slowing or retarding the movement of a solute, and is defined for solutes subject to equilibrium reactions with the soil matrix.

retentivity profile, soil–A graph showing the retaining capacity of a soil as a function of depth. The retaining capacity may be for water, for water at any given tension, for cations, or for any other substances held by soils.

reticulate mottling–A network of **mottles** with no dominant color, most commonly found in deeper horizons of soils containing **plinthite**.

rhizobia–Bacteria able to live symbiotically in roots of leguminous plants, from which they receive energy and often utilize molecular nitrogen. Collective common name for the genus *Rhizobium*.

rhizobia free–Any material that does not contain rhizobia able to nodulate leguminous plants of interest. The material need not be void of all rhizobia. See also **rhizobia populated**.

rhizobia populated–Any material that contains rhizobia able to nodulate leguminous plants of interest. Contrast with **rhizobia free**.

rhizocylinder–The plant root plus the adjacent soil that is influenced by the root. See also **rhizosphere**.

rhizoplane–Plant root surfaces usually including the adhering soil particles.

rhizosphere–The zone of soil immediately adjacent to plant roots in which the kinds, numbers, or activities of microorganisms differ from that of the bulk soil.

ridge–See **tillage**, *ridge*.

ridge planting–See **tillage**, *ridge planting*.

rill–See **erosion**, *rill*.

riparian–Land adjacent to a body of water that is at least periodically influenced by flooding. See also **flood plain, tidal flats**, and **wetland**.

river wash–In soil survey a map unit that is a miscellaneous area, which is barren **alluvial** areas of unstablilized sand silt, clay or gravel reworked by frequently by stream activity. See also **miscellaneous area**.

rock fragments–Unattached pieces of rock 2 mm in diameter or larger that are strongly cemented or more resistant to rupture. See **Table 3** for terms that are used to classify rock fragments in soils.

rock land–Areas containing frequent rock outcrops and shallow soils. Rock outcrops usually occupy from 25 to 90% of the area. See also **miscellaneous area**.

rock outcrop–In soil survey a map unit that is a miscellaneous area, which consists of exposures of bedrock other than lava flows and rock-lined pits. See also **miscellaneous area**.

rod weeding–See **tillage**, *rod weeding.*

rolling–See **tillage**, *rolling.*

root bed–See **tillage**, *root bed.*

rotary hoeing–See **tillage**, *rotary hoeing.*

rotary tilling–See **tillage**, *rotary tilling.*

rough broken land–Areas with very steep topography and numerous intermittent drainage channels but usually covered with vegetation. See also **miscellaneous areas** and **badlands**. (Not used in current U.S. system of soil taxonomy.)

r-selected–In ecological theory, that group of organisms in soil that rapidly proliferate in response to an abundance of resources. Analogous to **zymogenous** microorganisms.

rubble land–Areas with 90% or more of the surface covered with cobbles, stones, and boulders. Commonly occurs as **colluvium** at the base of mountains but some areas may be left on mountainsides by **glaciation** or **periglacial** processes. A **miscellaneous area**.

runoff–That portion of precipitation or irrigation on an area which does not infiltrate, but instead is discharged from the area. That which is lost without entering the soil is called *surface runoff.* That which enters the soil before reaching a stream channel is called *ground water runoff or seepage flow* from ground water. (In soil science *runoff* usually refers to the water lost by surface flow; in geology and hydraulics *runoff* usually includes both surface and subsurface flow.)

Table 3. Classification of rock fragments †

Shape and Size	Noun	Adjective
Spherical, cubelike, or equiaxial		
2-75 mm diameter	Pebbles	Gravelly
2-5 mm diameter	Fine	Fine gravelly
5-20 mm diameter	Medium	Medium gravelly
20-75 mm diameter	Coarse	Coarse gravelly
75-250 mm diameter	Cobbles	Cobbly
250-600 mm diameter	Stones	Stony
>600 mm diameter	Boulders	Bouldery
Flat		
2-150 mm long	Channers	Channery
150-380 mm long	Flagstones	Flaggy
380-600 mm long	Stones	Stony
>600 mm long	Boulders	Bouldery

† From: Soil survey division staff. 1993. *Soil survey manual*, USDA-SCS Agric. Handb. 18. p. 143, U.S. Gov. Print. Office, Washington, DC.

S

salic horizon–A mineral soil horizon of enrichment with secondary salts more soluble in cold water than gypsum. A salic horizon is 15 cm or more in thickness, contains at least 20 g kg^{-1} salt, and the product of the thickness in centimeters and amount of salt by weight is >600 g kg^{-1}.

Salids–**Aridisols** which have a **salic horizon** that has its upper boundary within 100 cm of the soil surface. (A suborder in the U.S. system of soil taxonomy.)

saline seep–Intermittent or continuous saline water discharge at or near the soil surface under dryland conditions which reduces or eliminates crop growth. It is differentiated from other saline soil conditions by recent and local origin, shallow water table, saturated root zone, and sensitivity to cropping systems and precipitation.

saline soil–A nonsodic soil containing sufficient soluble salt to adversely affect the growth of most crop plants. The lower limit of saturation extract electrical conductivity of such soils is conventionally set at 4 dS m^{-1} (at 25°C). Actually, sensitive plants are affected at half this salinity and highly tolerant ones at about twice this salinity.

saline-alkali soil–(no longer used in SSSA publications) (i) A soil containing sufficient exchangeable sodium to interfere with the growth of most crop plants and containing appreciable quantities of soluble salts. The exchangeable-sodium percentage is >15, the conductivity of the saturation extract >4 dS m^{-1} (at 25°C), and the pH is usually 8.5 or less in the saturated soil. (ii) A saline-alkali soil has a combination of harmful qualities of salts and either a high alkalinity or high content of exchangeable sodium, or both, so distributed in the profile that the growth of most crop plants is reduced. See also **saline-sodic soil**.

saline-sodic soil–(no longer used in SSSA publications) A soil containing sufficient exchangeable sodium to interfere with the growth of most crop plants and containing appreciable quantities of soluble salts. The exchangeable sodium ratio is greater than 0.15, conductivity of the soil solution, at saturated water content, of >4dS m^{-1} (at 25°C), and the pH is usually 8.5 or less in the saturated soil. See also **saline-alkali soil**.

salinity, soil–The amount of soluble salts in a soil. The conventional measure of soil salinity is the **electrical conductivity** of a saturation extract.

salinization–The process whereby soluble salts accumulate in the soil.

salt-affected soil–Soil that has been adversely modified for the growth of most crop plants by the presence of soluble salts, with or without high amounts of exchangeable sodium. See also **saline soil**, **saline-sodic soil**, and **sodic soil**.

salt balance–The quantity of soluble salt removed from an irrigated area in the drainage water minus that delivered in the irrigation water.

salt flats–In Soil Survey a map unit that is a miscellaneous area, composed of undrained flats in arid regions that have surface deposits of secondary salt overlying stratified and strongly saline sediment. See also **miscellaneous area**.

salt tolerance–The ability of plants to resist the adverse, nonspecific effects of excessive soluble salts in the rooting medium.

saltation–See **erosion**, *saltation*.

saltation flux–See **erosion**, *saltation flux*.

sample–A part of a population taken to estimate a parameter of the whole population.

sample plot–An area of land, usually small, used for measuring or observing performance under existing or applied treatments.

sand–(i) A soil separate. See also **soil separates**. (ii) A soil textural class. See also **soil texture**.

sand sheet–A large, irregularly shaped, commonly thin, surficial mantle of eolian sand, lacking the discernible slip faces that are common on dunes.

sandy–(i) Texture group consisting of sand and loamy sand textures. See also **soil texture**. (ii) Family particle-size class for soils with sand or loamy sand textures and <35% rock fragments in upper subsoil horizons.

sandy clay–A soil textural class. See also **soil texture**.

sandy clay loam–A soil textural class. See also **soil texture**.

sandy loam–A soil textural class. See also **soil texture**.

sapric material–Organic soil material that contains less than 1/6 recognizable fibers (after rubbing) of undecomposed plant remains. Bulk density is usually very low, and water holding capacity very high.

Saprists–**Histosols** that have a high content of plant materials so decomposed that original plant structures cannot be determined and a bulk density of about 0.2 Mg m^{-3} or more. **Saprists** are saturated with water for periods long enough to limit their use for most crops unless they are artificially drained. (A suborder in the U.S. system of soil taxonomy.)

saprolite–Soft, friable, isovolumetrically weathered bedrock that retains the fabric and structure of the parent rock exhibiting extensive inter-crystal and intra-crystal weathering. In pedology, saprolite was formerly applied to any unconsolidated residual material underlying the soil and grading to hard bedrock below.

saprophyte–An organism that lives on dead organic material.

saprophytic competence–The ability of a nodule symbiont or pathogenic microorganism to establish itself and live in soil as a saprophyte.

sate (synonym of satiate)–To fill most of the pores between soil particles with liquid, the lack of complete filling being caused by the entrapment of air as water enters the soil.

saturate–(i) To fill all the voids between soil particles with a liquid. (ii) To form the most concentrated solution possible under a given set of physical conditions in the presence of an excess of the solute. (iii) To fill to capacity, as the adsorption complex with a cation species; e.g., H$^+$-saturated, etc.

saturated soil paste–A particular mixture of soil and water. At saturation, the soil paste glistens as it reflects light, flows slightly when the container is tipped, and the paste slides freely and cleanly from a spatula.

saturation content–The mass water content of a saturated soil paste.

saturation extract–The solution extracted from a soil at its saturation water content.

scalping–A method of preparing forest soils for planting or seeding that consists of removing the ground vegetation and root mat to expose mineral soil.

scarifying–See **tillage**, *scarifying*.

scarp–An **escarpment**, cliff, or steep slope of some extent along the margin of a plateau, mesa, terrace, or structural bench. A scarp may be of any height.

scoria land–In soil survey a map unit that is a **miscellaneous area**, which consists of clinkers, burned shale or fine-grained sandstone remaining after coal beds burn out. See also **miscellaneous areas**.

screefing–A method of preparing forest soils for planting or seeding that consists of mechanically pushing aside the humus layer to expose mineral soil.

screen–See **irrigation.**

seal–See **surface sealing**.

second bottom–The first **terrace** above the normal flood plain of a stream.

secondary metabolite–A product of intermediary metabolism released from a cell.

secondary mineral–A mineral resulting from the decomposition of a primary mineral or from the reprecipitation of the products of decomposition of a primary mineral. See also **primary mineral**.

secondary nutrients–Refers to Ca, Mg, and S in fertilizers.

sediment–Transported and deposited particles or aggregates derived from rocks, soil, or biological material.

sedimentary rock–A rock formed from materials deposited from suspension or precipitated from solution and usually being more or less consolidated. The principal sedimentary rocks are sandstones, shales, limestones, and conglomerates.

sedimentation–The process of sediment deposition.

sedimentology–The science dealing with the study of processes of sedimentation and sediment properties.

seedbed–See **tillage**, *seedbed.*

segregated ice–Massive ice in a soil pedon, which is relatively free of soil particles.

selective cutting–(forestry) A system of cutting in which trees, usually the largest, or small groups of such trees are removed for commercial production or to encourage reproduction under the remaining stand in the openings.

selective enrichment–A technique for specifically encouraging the growth of a particular organism or group of organisms. See also **enrichment culture**.

selectivity coefficient–A conditional equilibrium coefficient for an ion exchange reaction that is expressed in terms of concentration variables for the exchangeable ions and either concentration variables or activities of the ions in solution.

self-mulching soil–A soil in which the surface layer becomes so well aggregated that it does not crust and seal under the impact of rain but instead serves as a surface mulch upon drying.

sensor–Any device which gathers electromagnetic radiation (EMR) or other energy and presents it in a form suitable for obtaining information about the environment. Passive sensors, such as thermal infrared and microwave, utilize EMR produced by the surface or object being sensed. Active sensors, such as radar, supply their own energy source. Aerial cameras use natural or artificially produced EMR external to the object or surface being sensed.

separate, soil–See **soil separate**.

sepiolite–$Si_{12}Mg_8O_{30}(OH)_4(OH_2) \cdot 8H_2O$ A fibrous clay mineral composed of two silica tetrahedral sheets and one magnesium octahedral sheet that make up the 2:1 layer. The 2:1 layers occur in strips with an average width of three linked tetrahedral chains joined at the edges to form tunnels where water molecules are held..

sequuum–(pl. sequa) A B horizon together with any overlying **eluvial** horizons.

series, soil–See **soil series**.

sesquan–A **cutan** composed of a concentration of **sesquioxides**.

sesquioxides–A general term for oxides and hydroxides of iron and aluminum.

shatter–See **tillage**, *shatter*.

shear–Force, as with a tillage tool, acting at a right angle to the direction of movement of the tillage implement.

shearing–See **tillage**, *shearing*.

shear strength–The maximum resistance of a soil to shearing stresses.

sheet erosion–See **erosion**.

sheet of polyhedra–See **phyllosilicate mineral terminology**.

shelter belt–See **erosion**, *windbreak*.

shoulder–The hillslope position that forms the uppermost inclined surface near the top of a slope. If present, it comprises the transition zone from backslope to summit. This position is dominantly convex in profile and erosional in origin.

shrinkage, soil –The process of soil material contracting to a lesser volume when subject to loss of water.

basic shrinkage phase (or zone)–The middle phase of soil shrinkage between the structural and residual shrinkage; it refers to the fundamental shrinkage process of a specified soil.

isotropic shrinkage–Shrinkage that occurs equally in all directions.

moisture ratio–Volume water per volume of soil [m^3 m^{-3}].

ped (shrinkage)–A naturally occurring unit of soil defined by surrounding lines of weakness; the smallest unit of natural soil with no internal shrinkage cracks.

residual shrinkage–Shrinkage that is less than volume water loss during the final stages of drying.

shrinkage characteristic–The relationship between the soil volume and volume of water contained in a specified soil mass or ped [m^3 m^{-3}].

shrinkage coefficient–The change in soil bulk volume with change in mass water content at a constant stress; also equivalent to, the rate of change in void ratio with moisture ratio at a constant stress.

structural shrinkage–Shrinkage that is less than volume water loss due to water drainage from macropores at high soil water content.

surface subsidence–See **shrinkage, soil**, *vertical shrinkage*.

swelling hysteresis–See **hysteresis**.

unidimensional shrinkage or 1-D shrinkage–Shrinkage that occurs exclusively in the vertical direction.

unitary shrinkage–Shrinkage which is equivalent to the change in water volume.

vertical shrinkage–The shrinkage-induced length change of a soil in the vertical direction, also called *surface subsidence* if it occurs exclusively at the soil surface.

shrub-coppice dune–A small, streamlined dune that forms around desert, brush-and-clump vegetation.

side slope–The slope bounding a drainageway and lying between the drainageway and the adjacent interfluve. It is generally linear along the slope width and overland flow is parallel down the slope. See also **nose slope**.

siderophore–A nonporphyrin metabolite secreted by certain microorganisms that forms a highly stable coordination compound with iron. There are two major types: catecholate and hydroxamate.

Sierozem–A **zonal** great soil group consisting of soils with pale grayish A horizons grading into calcareous material at a depth of 1 foot or less, and formed in temperate to cool, arid climates under a vegetation of desert plants, short grass, and scattered brush. (Not used in current U.S. system of soil taxonomy.)

significant–(statistics) A term applied to differences, correlation, etc., to indicate that they are probably not due to chance alone; usually indicates a probability of not less than 95 percent.

silica-alumina ratio–The molecules of silicon dioxide (SiO_2) per molecule of aluminum oxide (Al_2O_3) in clay minerals or in soils.

silica-sesquioxide ratio–The molecules of silicon dioxide (SiO_2) per molecule of aluminum oxide (Al_2O_3) plus ferric oxide (Fe_2O_3) in clay minerals or in soils.

silt–(i) A soil separate. See also **soil separates**. (ii) A soil textural class. See also **soil texture**.

silt loam–A soil textural class. See also **soil texture**.

silting–The deposition of silt from a body of standing water; choking, filling, or covering by stream-deposited silt that occurs in a place of retarded flow or behind a dam or reservoir. The term often includes particles from clay to sand-size.

silty clay–A soil textural class. See also **soil texture**.

silty clay loam–A soil textural class. See also **soil texture**.

sinkhole–A closed depression formed either by solution of the surficial bedrock (e.g., limestone, gypsum, or salt) or by collapse of underlying caves. Complexes of sinkholes in carbonate-rock terrain are the main components of **karst** topography.

siphon tubes–See **irrigation,** *siphon tubes*.

site–A volume defined by the abiotic factors (i.e. climate, soil, physiography) that influence vegetation growth and development.

site index–The height of the dominant and codominant trees (not suppressed during development) at an index age, commonly 25, 50, or 100 years. Used in conjunction with volume tables, site index provides an indication of relative site production.

site productivity–The capacity of a site to produce specific products (i.e. biomass or lumber volume) for a given vegetative configuration over time as influenced by abiotic factors (i.e. soil, climate, physiography). **Net primary productivity (NPP)** provides the fundamental measure of site productivity. When measured at the point of leaf carrying capacity for all potential flora, NPP is a measure of potential site productivity. Rate of product growth, an economic component, is occasionally used as a partial measure of site productivity.

site quality–A relative measure of the vegetative production capacity of a site for a given purpose.

skeletan–A **cutan** composed of **skeleton grains**.

skeleton grains–Individual grains that are relatively stable and not readily translocated, concentrated, or reorganized by soil-forming processes; they include mineral grains and resistant siliceous and organic bodies larger than colloidal size.

skew planes–Planar voids that traverse the soil material in an irregular manner, having no specific distribution or orientation pattern between individuals.

slickens–Accumulations of fine-textured material, such as separated in placer mining and in ore mill operations; may be detrimental to plant growth and are usually confined in specially constructed basins. A also **miscellaneous area**.

slickensides–Stress surfaces that are polished and striated produced by one mass sliding past another. **Slickensides** are common below 50 cm in swelling clays subject to large changes in water content.

slick spots–Areas having a puddled or crusted, very smooth, nearly impervious surface. The underlying material is dense and massive. A also **miscellaneous area**.

slit planting–See **tillage**, *slit planting*.

slot planting–See **tillage**, *slit planting*.

slough–(i) A **swamp** or shallow lake system in northern and midwestern USA. (ii) A slowly flowing shallow **swamp** or **marsh** in southeastern USA.

slow release–A fertilizer term used interchangeably with *delayed release*, *controlled release*, *controlled availability*, *slow acting*, and *metered release* to designate a rate of dissolution (usually in water) much less than is obtained for completely water-soluble compounds. Slow release may involve either compounds that dissolve slowly or soluble compounds coated with substances relatively impermeable to water.

slump–(i) A mass movement process characterized by a **landslide** involving a shearing and rotary movement of a generally independent mass of rock or earth along a curved slip surface (concave upward) and about an axis parallel to the slope from which it descends, and by backward tilting of the mass with respect to that slope so that the slump surface often exhibits a reversed slope facing uphill. (ii) The landform or mass of material slipped down during, or produced by a slump.

slump block–The mass of material torn away as a coherent unit during slumping.

s-matrix (of a soil material)–The material within the simplest peds, or composing **apedal soil materials**, in which the pedological features occur; it consists of the **plasma**, **skeleton grains** and **voids** that do not occur as pedological features other than those expressed by specific extinction (orientation) patterns. Pedological features also have an internal s-matrix.

smectite–A group of 2:1 layer silicates with a high cation exchange capacity, about 110 $cmol_c$ kg^{-1} for soil smectites, and variable interlayer spacing. Formerly called the montmorillonite group. The group includes dioctahedral members **montmorillonite**, beidellite, and nontronite, and trioctahedral members saponite, hectorite, and sauconite. See also **Appendix I, Table A3**.

sod planting–See **tillage**, *sod planting*.

sodic soil–A nonsaline soil containing sufficient exchangeable sodium to adversely affect crop production and soil structure under most conditions of soil and plant type. The sodium adsorption ratio of the saturation extract is at least 13.

sodication–The process whereby the exchangeable sodium content of a soil is increased.

sodium adsorption ratio (SAR)–A relation between soluble sodium and soluble divalent cations which can be used to predict the **exchangeable sodium** fraction of soil equilibrated with a given solution. It is defined as follows:

$$SAR = \frac{[\text{sodium}]}{[\text{calcium} + \text{magnesium}]^{1/2}}$$

where concentrations, denoted by brackets, are expressed in mmoles per liter.

sodium adsorption ratio, adjusted–The sodium adsorption ratio of a water adjusted for the precipitation or dissolution of Ca^{2+} that is expected to occur where a water reacts with alkaline earth carbonates within a soil.

soil–(i) The unconsolidated mineral or organic material on the immediate surface of the earth that serves as a natural medium for the growth of land plants. (ii) The unconsolidated mineral or organic matter on the surface of the earth that has been subjected to and shows effects of genetic and environmental factors of: climate (including water and temperature effects), and macro- and microorganisms, conditioned by relief, acting on parent material over a period of time. A product-soil differs from the material from which it is derived in many physical, chemical, biological, and morphological properties and characteristics.

soil aeration–The condition, and sum of all processes affecting, soil pore-space gaseous composition, particularly with respect to the amount and availability of oxygen for use by soil biota and/or soil chemical oxidation reactions.

soil air–The soil atmosphere; the gaseous phase of the soil, being that volume not occupied by solid or liquid.

soil amendment–Any material such as lime, gypsum, sawdust, compost, animal manures, crop residue or synthetic soil conditioners that is worked into the soil or applied on the surface to enhance plant growth. Amendments may contain important **fertilizer** elements but the term commonly refers to added materials other than those used primarily as **fertilizers**. See also **soil conditioner**.

soil association–A kind of map unit used in soil surveys comprised of **delineations**, each of which shows the size, shape, and location of a landscape unit composed of two or more kinds of component soils or component soils and miscellaneous areas, plus allowable inclusions in either case. The individual bodies of component soils and miscellaneous areas are large enough to be delineated at the scale of 1:24 000. Several to numerous bodies of each kind of component soil or miscellaneous area are apt to occur in each delineation and they occur in a fairly repetitive and describable pattern. See also **component soil**, **soil consociation**, **undifferentiated group**, **miscellaneous areas**.

soil auger–A tool for boring into the soil and withdrawing a small sample for field or laboratory observation. Soil augers may be classified into several types as follows: (i) those with worm-type bits, unenclosed; (ii) those with worm-type bits enclosed in a hollow cylinder; and (iii) those with a hollow cylinder with a cutting edge at the lower end.

soil biochemistry–The branch of soil science concerned with enzymes and the reactions, activities, and products of soil microorganisms.

soil characteristics–Soil properties which can be described or measured by field or laboratory observations, e.g., color, temperature, water content, structure, pH, and exchangeable cations.

soil chemistry–The branch of soil science that deals with the chemical constitution, chemical properties, and chemical reactions of soils.

soil classification–See **classification, soil**.

soil compaction–Increasing the soil **bulk density**, and concomitantly decreasing the soil porosity, by the application of mechanical forces to the soil.

soil complex–A kind of map unit used in soil surveys comprised of **delineations**, each of which shows the size, shape and location of a landscape unit composed of two or more kinds of component soils, or component soils and a miscellaneous area, plus allowable inclusions in either case. The individual bodies of component soils and miscellaneous areas are too small to be delineated at the scale of 1:24 000. Several to numerous bodies of each kind of component soil or the miscellaneous area are apt to occur in each delineation. See also **component soil, soil consociation, soil association, undifferentiated group, miscellaneous areas**.

soil conditioner–A material which measurably improves specific soil physical characteristics or physical processes for a given use or as a plant growth medium. Examples include sawdust, peat, compost, synthetic polymers, and various inert materials. See also **soil amendment**.

soil conservation–(i) Protection of the soil against physical loss by erosion or against chemical deterioration; that is, excessive loss of fertility by either natural or artificial means. (ii) A combination of all management and land use methods that safeguard the soil against depletion or deterioration by natural or by human-induced factors. (iii) The branch of soil science that deals with soil conservation (i) and (ii).

soil consociation–A kind of map unit comprised of **delineations**, each of which shows the size, shape, and location of a landscape unit composed of one kind of component soil, or one kind of miscellaneous area, plus allowable inclusions in either case. See also **component soil, soil complex, soil association, undifferentiated group, miscellaneous areas**.

soil creep–See **creep**.

soil extract–The solution separated from a soil suspension or from a soil by filtration, centrifugation, suction, or pressure. (May or may not be heated prior to separation.)

soil fabric–The combined influence of the shape, size, and spatial arrangement of soil solids and soil pores.

soil fertility–The quality of a soil that enables it to provide nutrients in adequate amounts and in proper balance for the growth of specified plants or crops.

soil formation factors–The variables, usually interrelated natural agencies, that are active in and responsible for the formation of soil. The factors are usually grouped into five major categories as follows: parent material, climate, organisms, topography, and time.

soil genesis–(i) The mode of origin of the soil with special reference to the processes or **soil-forming factors** responsible for development of the **solum**, or true soil, from unconsolidated parent material. (ii) The branch of soil science that deals with soil genesis.

soil geography–The branch of physical geography that deals with the areal distributions of soils.

soil heat-flux density–The amount of heat entering a specified cross-sectional area of soil per unit time.

soil horizon–A layer of soil or soil material approximately parallel to the land surface and differing from adjacent genetically related layers in physical, chemical, and biological properties or characteristics such as color, structure, texture, consistency, kinds and number of organisms present, degree of acidity or alkalinity, etc. See also **Appendix II**.

soil hydrophobicity–The tendency for a soil particle or soil mass to resist hydration, usually quantified using the water drop penetration time test. See also **soil wettability, water drop penetration time**.

soil interpretations–Predictions of soil behavior in response to specific uses or management based on inferences from soil characteristics and qualities (e.g., trafficability, erodibility, productivity, etc.). They are either qualitative or quantitative estimates or ratings of soil productivities, potentials, or limitations.

soil loss tolerance (T value)–See **erosion**, *soil loss tolerance (T value)*.

soil management–(i) The sum total of all tillage and planting operations, cropping practices, fertilizer, lime, irrigation, herbicide and insecticide application, and other treatments conducted on or applied to a soil for the production of plants. See also **tillage**, *soil management*. (ii) The branch of soil science that deals with the items listed in (i).

soil management groups–Groups of taxonomic soil units with similar adaptations or management requirements for one or more specific purposes, such as: adapted crops or crop rotations, drainage practices, fertilization, forestry, highway engineering, etc.

soil map–A map showing the distribution of soils or other soil map units in relation to the prominent physical and cultural features of the earth's surface. The following kinds of soil maps are recognized in the USA:

> *soil map, detailed*–A soil map on which the boundaries are shown between all soils that are significant to potential use as field management systems. The scale of the map will depend upon the purpose to be served, the intensity of land use, the pattern of soils, and the scale of the other cartographic materials available. Traverses are usually made at 400-m, or more frequent, intervals. Commonly a scale of 10 cm = 1609 m is now used for field mapping in the USA.

> *soil map, detailed reconnaissance*–A reconnaissance map on which some areas or features are shown in greater detail than usual, or than others.

> *soil map, generalized*–A small-scale soil map which shows the general distribution of soils within a large area and thus in less detail than on a detailed soil map. Generalized soil maps may vary from soil association maps of a county, on a scale of 1 cm = 633 m, to maps of larger regions showing associations dominated by one or more great soil groups.

> *soil map, reconnaissance*–A map showing the distribution of soils over a large area as determined by traversing the area at intervals varying from about 800 m to several kilometers. The units shown are soil associations. Such a map is usually made only for exploratory purposes to outline areas of soil suitable for more intensive development. The scale is usually much smaller than for detailed soil maps.

soil map, schematic–A soil map compiled from scant knowledge of the soils of new and undeveloped regions by the application of available information about the soil-formation factors of the area. Usually on a small scale (1:1 000 000 or smaller).

soil mechanics and engineering–The branches of engineering and soil science that deal with the effect of forces on the soil and the application of engineering principles to problems involving the soil.

soil microbiology–The branch of soil science concerned with soil-inhabiting microorganisms, their functions, and activities.

soil micromorphology–The study of soil morphology by microscopic (light optical and less frequently by submicroscopic) methods, often using thin-section techniques.

soil mineral–(i) Any mineral that occurs as a part of or in the soil. (ii) A natural inorganic compound with definite physical, chemical, and crystalline properties (within the limits of isomorphism), that occurs in the soil. See also **clay mineral**.

soil mineralogy–The branch of soil science that deals with the homogeneous inorganic materials found in the earth's crust to the depth of weathering or of sedimentation.

soil moisture regimes–See **aquic, aridic, torric, udic, ustic, xeric**.

soil-moisture tension–See **soil water**, *soil water potential*.

soil monolith–A vertical section of a soil profile removed from the soil and mounted for display or study.

soil morphology–(i) The physical constitution of a soil profile as exhibited by the kinds, thickness, and arrangement of the horizons in the profile, and by the texture, structure, consistence, and porosity of each horizon. (ii) The visible characteristics of the soil or any of its parts.

soil order–A group of soils in the broadest category. For example, in the 1938 classification system. The three soil orders were zonal soil, intrazonal soil, and azonal soil. In the 1975 there were 10 orders, whereas in the current USDA classification scheme (Soil Survey Staff. 1994. *Soil taxonomy*: A basic system of soil classification for making and interpreting soil surveys. SCS-USDA. U.S. Gov. Print. Office, Washington, DC) there are 11 orders, differentiated by the presence or absence of diagnostic horizons: **Alfisols, Andisols, Aridisols, Entisols, Histosols, Inceptisols, Mollisols, Oxisols, Spodosols, Ultisols, Vertisols**. Orders are divided into Suborders and the Suborders are farther divided into Great Groups.

soil organic matter–The organic fraction of the soil exclusive of undecayed plant and animal residues. See also **humus**.

soil organic residue–Animal and vegetative materials added to the soil of recognizable origin.

soil oxygen diffusion rate–(i) The rate of diffusion of oxygen through soil as defined by Fick's law. (ii) A measurement of diffusion governed oxygen reduction rate at the surface of platinum microelectrodes used to assess the oxygen supplying ability of the soil relative to the needs of plant roots, usually referred to as soil ODR.

soil physics–The branch of soil science that deals with the physical properties of the soil, with emphasis on the state and transport of matter (especially water) and energy in the soil.

soil piping or **tunneling**–Accelerated erosion which results in subterranean voids and tunnels.

soil population–(i) All the organisms living in the soil, including plants and animals. (ii) Members of the same taxa. (iii) Delineations of the same map unit – a grouping of like things in a statistical sense.

soil pores–That part of the bulk volume of soil not occupied by soil particles. Soil pores have also been referred to as interstices or voids.

soil productivity–The capacity of a soil to produce a certain yield of crops or other plants with a specified system of management.

soil qualities–Inherent attributes of soils which are inferred from soil characteristics or indirect observations (e.g., compactibility, erodibility, and fertility).

soil quality–The capacity of a soil to function within ecosystem boundaries to sustain biological productivity, maintain environmental quality, and promote plant and animal health.

soil sample–A representative sample taken from an area, a field, or portion of a field from which the physical, biological, and chemical properties can be determined.

soil science–That science dealing with soils as a natural resource on the surface of the earth including soil formation, classification and mapping; physical, chemical, biological, and fertility properties of soils per se; and these properties in relation to the use and management of soils.

soil separates–Mineral particles, <2.0 mm in equivalent diameter, ranging between specified size limits. The names and size limits of separates recognized in the USA are: *very coarse sand*,[1] 2.0 to 1.0 mm; *coarse sand,* 1.0 to 0.5 mm; *medium sand,* 0.5 to 0.25 *mm; fine sand,* 0.25 to 0.10 mm; *very fine sand,* 0.10 to 0.05 mm; *silt,* 0.05 to 0.002 mm; and *clay,*[2] *<0.002 mm.*

The separates recognized by the International Society of Soil Science are: (i) coarse sand, 2.0 to 0.2 mm; (ii) *fine sand,* 0.2 to 0.02 mm; (iii) *silt,* 0.02 to 0.002 mm; and (iv) *clay,* <0.002 mm.

soil series–The lowest category of U.S. system of soil taxonomy; a conceptualized class of soil bodies (polypedons) that have limits and ranges more restrictive than all higher taxa. Soil series are commonly used to name dominant or codominant polypedons represented on detailed soil maps. The soil series serve as a major vehicle to transfer soil information and research knowledge from one soil area to another.

soil solution–The aqueous liquid phase of the soil and its solutes.

soil strength (cone index, penetration resistance)–A transient localized soil property which is a combined measure of a given pedon's, horizon's, or other soil subunit's solid phase adhesive and cohesive status. This property is most easily affected by changes in soil water content and bulk density, although other factors including texture, mineralogy, cementation, cation composition and organic matter content also affect it. In situ characterization with soil

---

[1]　Prior to 1947 this separate was called "fine gravel;" now fine gravel includes particles between 2.0 mm and about 12.5 mm in diameter.

[2]　Prior to 1937, "clay" included particles <0.005 mm in diameter, and "silt," those particles from 0.05 to 0.005 mm.

penetrometer is the most common agricultural measure of soil strength, although measurements of other engineering components of strength on disturbed samples are also regarded as valid characterizations.

soil structure–The combination or arrangement of primary soil particles into secondary units or **peds**. The secondary units are characterized on the basis of size, shape, and grade (degree of distinctness). See also **soil structure grades** and **soil structure shapes, Table 4**.

soil structure grades–A grouping or classification of soil structure on the basis of inter- and intra-aggregate adhesion, cohesion, or stability. Four grades of structure are recognized as follows:

> *Structureless*–No observable aggregation or no definite and orderly arrangement of natural lines of weakness. *Massive,* if coherent; *single-grain,* if noncoherent.

> *Weak*–Poorly formed indistinct peds, barely observable in place. When gently disturbed, the soil material parts into a mixture of whole and broken units and much material that exhibits no planes of weakness.

> *Moderate*–Well-formed distinct peds evident in undisturbed soil. When disturbed, soil material parts into a mixture of whole units, broken units, and material that is not in units.

> *Strong*–Peds are distinct in undisturbed soil. They separate cleanly when soil is disturbed, and the soil material separates mainly into whole units when removed.

soil structure shapes–A classification of soil structure based on the shape of the aggregates or peds in the profile. See also **soil structure** and **Table 4**.

soil structure sizes–See **soil structure** and **Table 4**.

soil surface seal–See **surface sealing.**

soil survey–(i)The systematic examination, description, classification, and mapping of soils in an area. Soil surveys are classified according to the kind and intensity of field examination. (ii) The program of the National Cooperative Soil Survey that includes developing and implementing standards for describing, classifying, mapping, writing, and publishing information about soils of a specific area.

Soil taxonomy–U.S. Department of Agriculture soil classification system.

soil temperature regimes–See **pergelic, cryic, frigid, mesic, thermic, hyperthermic**.

soil test–A chemical, physical, or biological procedure that estimates the suitability of the soil to support plant growth. (Sometimes used as an adjective to define fractions of soil components, e.g., "soil test phosphorus".)

soil test calibration–The process of determining the crop nutrient requirement at different soil test values.

soil test correlation–The process of determining the relationship between plant nutrient uptake or yield and the amount of nutrient extracted by a particular soil test method.

soil test critical concentration–The concentration of an extractable nutrient above which a crop response to added nutrient would not be expected.

soil test interpretation–The process of developing nutrient application recommendations from soil test concentrations, and other soil, crop, economic, environmental, and climatic information.

Table 4—Shapes and size classes of soil structure†

		Shape of structure			
Units are flat and platelike. They are generally oriented horizontally and faces are mostly horizontal	Units are prismlike and bounded by flat to rounded vertical faces. Units are distinctly longer vertically than horizontally; vertices angular.		Units are blocklike or polyhedral with flat or slightly rounded surfaces that are casts of the faces of surrounding peds; nearly equidimensional		Units are approximately spherical or polyhedral and are bounded by curved or very irregular faces that are not casts of adjoining peds
	Tops of units are indistinct and normally flat.	Tops of units are very distinct and normally rounded.	Faces intersect at relatively sharp angles	Mixture of rounded and plane faces and the vertices are mostly rounded	
Platy	Prismatic	Columnar	Angular blocky	Subangular blocky	Granular
Size class					
	mm	mm	mm	mm	mm	mm
Very fine or very thin‡	<1	<10	<10	<5	<5	<1
Fine or thin‡	1-2	10-20	10-20	5-10	5-10	1-2
Medium	2-5	20-50	20-50	10-20	10-20	2-5
Coarse or thick‡	5-10	50-100	50-100	20-50	20-50	5-10
Very coarse or very thick‡	>10	>100	>100	>50	>50	>10

† From: Soil survey division staff. 1993. *Soil survey manual*, USDA-SCS Agric. Handb. 18. U.S. Gov. Print. Office, Washington, DC.
‡ In describing plates, *thin* is used instead of *fine* and *thick* is used instead of *coarse*.

soil texture–The relative proportions of the various **soil separates** in a soil as described by the classes of soil texture shown in Fig. 1. The textural classes may be modified by the addition of suitable adjectives when rock fragments are present in substantial amounts; for example, "stony silt loam." (For other modifiers see also **rock fragments**.) The **sand**, **loamy sand**, and **sandy loam** are further subdivided on the basis of the proportions of the various sand separates present. The limits of the various classes and subclasses are as follows:

clay–Soil material that contains 40% or more clay, <45% sand, and <40% silt.

clay loam–Soil material that contains 27 to 40% clay and 20 to 45% sand.

loam–Soil material that contains 7 to 27% clay, 28 to 50% silt, and <52% sand.

loamy sand–Soil material that contains between 70 and 91% sand and the percentage of silt plus 1.5 times the percentage of clay is 15 or more; and the percentage of silt plus twice the percentage of clay is less than 30.

   loamy coarse sand– Soil material that contains 25% or more very coarse and coarse sand, and <50% any other one grade of sand.

   loamy sand– Soil material that contains 25% or more very coarse, coarse, and medium sand, <25% very coarse and coarse sand, and <50% fine or very fine sand.

   loamy fine sand– Soil material that contains 50% or more fine sand (or) <25% very coarse, coarse, and medium sand and <50% very fine sand.

   loamy very fine sand– Soil material that contains 50% or more very fine sand.

sand–Soil material that contains 85% or more of sand; percentage of silt, plus 1.5 times the percentage of clay, shall not exceed 15.

   coarse sand– Soil material that contains 25% or more very coarse and coarse sand, and <50% any other one grade of sand.

   sand– Soil material that contains 25% or more very coarse, coarse, and medium sand, <25% very coarse and coarse sand, and <50% fine or very fine sand.

   fine sand– Soil material that contains 50% or more fine sand (or) <25% very coarse, coarse, and medium sand and <50% very fine sand.

   very fine sand–Soil material that contains 50% or more very fine sand.

sandy clay–Soil material that contains 35% or more clay and 45% or more sand.

sandy clay loam–Soil material that contains 20 to 35% clay, <28% silt, and >45% sand.

sandy loam–Soil material that contains 7 to 20% clay, more than 52% sand, and the percentage of silt plus twice the percentage of clay is 30 or more; or less than 7% clay, less than 50% silt, and more than 43% sand.

   coarse sandy loam–Soil material that contains 25% or more very coarse and coarse sand and <50% any other one grade of sand.

   sandy loam–Soil material that contains 30% or more very coarse, coarse, and medium sand, but <25% very coarse and coarse sand, and <30% very fine or fine sand, or <15% very coarse, coarse, and medium sand and <30% either fine sand or very fine sand and 40% or less fine plus very fine sand.

fine sandy loam–Soil material that contains 30% or more fine sand and
<30% very fine sand (or) between 15 and 30% very coarse, coarse, and
medium sand, or >40% fine and very fine sand, at least half of which is
fine sand, and <15% very coarse, coarse, and medium sand.

very fine sandy loam–Soil material that contains 30% or more very fine sand
and <15% very coarse, coarse, and medium sand (or) >40% fine and
very fine sand, more than half of which is very fine sand and <15% very
coarse, coarse, and medium sand.

silt–Soil material that contains 80% or more silt and <12% clay.

silty clay–Soil material that contains 40% or more clay and 40% or more silt.

silty clay loam–Soil material that contains 27 to 40% clay and <20% sand.

silt loam–Soil material that contains 50% or more silt and 12 to 27% clay (or)
50 to 80% silt and <12% clay.

soil type–Formerly in the U.S. soil classification systems prior to publication of
USDA *Soil Taxonomy* (1975). (i) The lowest unit in the natural system of soil
classification; a subdivision of a soil series and consisting of or describing soils
that are alike in all characteristics including the texture of the A horizon or plow
layer; (ii) In Europe, roughly equivalent to a great soil group. See also **soil
series**.

soil variant–A soil whose properties are believed to be sufficiently different from
other known soils to justify a new series name but comprising such a limited
geographic area that creation of a new series is not justified. Use of this term
was discontinued in 1988. See also **taxadjunct**.

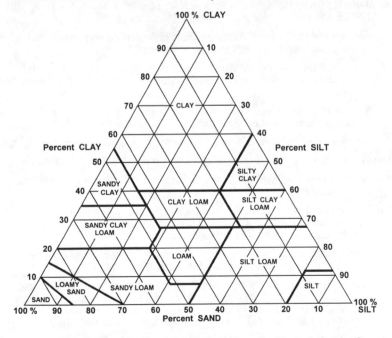

**Fig. 1.   Graph showing the percentages of sand, silt, and clay in the
soil texture classes.**

soil water –

soil water potential, (pressure, head)–The amount of work that must be done per unit of a specified quantity of pure water in order to transport reversibly and isothermally an infinitesimal quantity of water from a specified source to a specified destination.

If the specified quantity is volume, the potential is referred to as pressure (Pa).

If the specified quantity is weight, the potential is referred to as head (m).

If the specified quantity is mass, the potential is the term used (J kg^{-1}).

hydraulic head, pressure, potential–The sum of gravitational, hydrostatic, and matric water potential, expressed as head, pressure, or potential.

differential water capacity–The absolute value of the rate of change of water content with soil water pressure. The water capacity at a given water content will depend on the particular desorption or adsorption curve employed. Distinction should be made between volumetric and specific water capacity.

water content–The water lost from the soil upon drying to constant mass at 105°C; expressed either as the mass of water per unit mass of dry soil or as the volume of water per unit bulk volume of soil.

Table 5. Soil water terminology.

Type	Source	Destination
Osmotic	Pool of pure water at specified elevation and atmospheric pressure	Pool identical to the source pool but containing soil solution
Gravitational	Pool of soil solution at specified elevation and atmospheric pressure	Pool identical to the source pool but at the elevation of the point under consideration
Matric (above water table)	Pool of soil solution at the elevation and external air pressure of the point under consideration	Soil water at the point under consideration (above water table)
Hydrostatic (below water table)		Soil water at the point under consideration (below water table)
Air	External air pressure (atmospheric) at the elevation of the point under consideration	Soil air at the point under consideration.
Total	Pool of pure water at specified elevation and atmospheric pressure	Soil water at the point under consideration

hydraulic conductivity–The proportionality factor in **Darcy's law** as applied to the viscous flow of water in soil, i.e., the flux of water per unit gradient of hydraulic potential. If conditions require that the viscosity of the fluid be divorced from the conductivity of the medium, it is convenient to define the permeability (intrinsic permeability has been used in some publications) of the soil as the conductivity divided by the fluidity of the fluid. For the purpose of solving the partial differential equation of the nonsteady-state flow in unsaturated soil it is often convenient to introduce a variable termed the **soil water diffusivity**.

soil water diffusivity–The **hydraulic conductivity** divided by the **differential water capacity** (care being taken to be consistent with units), or the flux of water per unit gradient of water content in the absence of other force fields.

soil water characteristic or **characteristic curve**–The relationship between the soil-water content (by mass or volume) and the soil-water matric potential. Also called the **water retention curve** or isotherm, and the **water release curve**.

soil water pressure–See **soil water**, *soil water potential*.

soil wettability–See **water drop penetration time; soil hydrophobicity.**

solid set sprinkler irrigation–See **irrigation**, *sprinkler irrigation system, solid set*.

solifluction–Slow, viscous downslope flow of water-saturated regolith. Rates of flow vary widely. The presence of frozen substrate or even freezing and thawing is not implied in the original definition. However, one component of solifluction can be creep of frozen ground. The term is commonly applied to processes operating in both seasonal frost and permafrost areas.

Solonchak–A great soil group of the **intrazonal** order and **halomorphic** suborder, consisting of soils with gray, thin, salty crust on the surface, and with fine granular mulch immediately below being underlain with grayish, friable, salty soil; formed under subhumid to arid, hot or cool climate, under conditions of poor drainage, and under a sparse growth of halophytic grasses, shrubs, and some trees. (Not used in current U.S. system of soil taxonomy.)

Solonetz–A great soil group of the **intrazonal** order and **halomorphic** suborder, consisting of soils with a very thin, friable, surface soil underlain by a dark, hard columnar layer usually highly alkaline; formed under subhumid to arid, hot to cool climates, under better drainage than Solonchaks, and under a native vegetation of halophytic plants. (Not used in current U.S. system of soil taxonomy.)

solum (plural: sola)–A set of horizons that are related through the same cycle of pedogenic processes; the A, E, and B horizons.

sombric horizon–A subsurface mineral horizon that is darker in color than the overlying horizon but that lacks the properties of a spodic horizon. Common in cool, moist soils of high altitude in tropical regions.

sorption–The removal of an ion or molecule from solution by **adsorption** and **absorption** . It is often used when the exact nature of the mechanism of removal is not known.

sorptivity–$S = I\ t^{-\frac{1}{2}}$ for horizontal infiltration of water, where I is cumulative infiltration and t is time. **Sorptivity** is dependent on initial and boundary conditions of soil water content among other factors.

spatial variability–The variation in soil properties (i) laterally across the landscape, or (ii) vertically downward through the soil.

specific activity–Number of enzyme activity units per mass of protein. Often expressed as micromoles of product formed per unit time per milligram of protein. Also used in radiochemistry to express the radioactivity per mass of material (radioactive + nonradioactive).

specific adsorption–The strong adsorption of ions or molecules on a surface. Specifically adsorbed materials are not readily removed by ion exchange.

specific surface–The solid-particle surface area (of a soil or porous medium) divided by the solid-particle mass or volume, expressed in $m^2 kg^{-1}$ or $m^2 m^{-3} = m^{-1}$, respectively.

specific water capacity–The change of soil-water mass content with change in soil-water **matric potential**.

splash erosion–See **erosion**, *splash erosion*.

spodic horizon–A mineral soil horizon that is characterized by the **illuvial** accumulation of **amorphous materials** composed of aluminum and organic carbon with or without iron. The **spodic horizon** has a certain minimum thickness, and a minimum quantity of extractable carbon plus iron plus aluminum in relation to its content of clay.

Spodosols–Mineral soils that have a **spodic horizon** or a **placic horizon** that overlies a **fragipan**. (An order in the U.S. system of soil taxonomy.)

spoil bank–Rock waste, banks, and dump depositions resulting from the excavation of ditches and strip mines.

spores–Specialized reproductive cell. Asexual spores germinate without uniting with other cells, whereas sexual spores of opposite mating types unite to form a zygote before germination occurs.

spray irrigation–See **irrigation**, *spray irrigation*.

sprinkler–See *irrigation, sprinkler*.

standard cone (ASAE standard cone)–The cone-shaped tip used at the insertion end of soil **penetrometer** probes, following design criteria prescribed by the ASAE standard. Briefly, a 30 degree stainless steel cone having a basal diameter of either 20.27 mm or 12.83 mm.

static penetrometer–A **penetrometer** which is pushed into the soil at a constant and slow rate of penetration.

stemflow–That portion of precipitation or irrigation water that is intercepted by plants and then flows down the stem to the ground.

sterilization–Rendering an object or substance free of viable microbes.

sticky point–(i) A condition of consistency at which the soil barely fails to stick to a foreign object. (ii) Specifically and numerically, the water mass content of a well-mixed kneaded soil that barely fails to adhere to a polished nickel or stainless steel surface when the shearing speed is 50 mm s^{-1}.

Stokes' law–The equation expressing the force of viscous resistance on a smooth, rigid sphere moving in a viscous fluid under standard temperature and pressure, namely

$$F = 3\pi\eta DV$$

where F is the force of viscous resistance, $\pi = 3.1416$, η is the fluid viscosity, D is the diameter of the sphere, and V is the velocity of fall (or movement). Applying Stokes' law to gravity sedimentation as used in particle-size analysis of soil by pipette or hydrometer methods, the resulting sedimentation equation is

$$V = 2gr^2 (d_1 - d_2)/9\eta$$

where g is the acceleration of gravity, r is the "equivalent" radius of a particle, d_1 is the soil-particle density, and d_2 is the fluid density. Stokes' law applied to centrifugation yields still another equation for V.

stone line–A sheet-like lag concentration of **coarse fragments** in surficial sediments. In cross section, the line may be marked only by scattered fragments or it may be a discrete layer of fragments. The fragments are more often pebbles or cobbles than stones. A **stone line** generally overlies material that was subject to weathering, soil formation, and erosion before deposition of the overlying material. Many **stone lines** seem to represent buried erosion pavements, originally formed by running water on the land surface and concurrently covered by surficial sediment.

stones–Rock or mineral fragments between 250 and 600 mm in diameter if rounded, and 380 to 600 mm if flat. See also **rock fragments**.

stoniness–Classes based on the relative proportion of stones at or near the soil surface. Used as a phase distinction in mapping soils. See also **rock fragments**.

stony–(i) A stoniness class in which there are enough stones at or near the soil surface to be a continuing nuisance during operations that the mix the surface layer, but they do not make most such operations impractical. (ii) Containing appreciable quantities of stones. See also **rock fragments**.

strath terrace–A type of **stream terrace**, formed as an erosional surface cut on bedrock and thinly mantled with stream deposits (alluvium).

stratified–Arranged in or composed of strata or layers.

straw mulching–The use of straw to create a surface mulch on all or part of the soil surface for soil or water conservation, for soil temperature management or for weed suppression. See also **furrow mulching, stubble mulch.**

stream order–An integer system applied to tributaries (stream segments) that documents their relative position within a drainage basin network as determined by the pattern of its confluences. The order of the drainage basin is determined by the highest integer. Several systems exist. In the Strahler system, the smallest unbranched tributaries are designated order 1; the confluence of two first-order streams produces a stream segment of order 2; the junction of two second-order streams produces a stream segment of order 3, etc.

stream terrace–One of a series of platforms in a stream valley, flanking and more or less parallel to the stream channel, originally formed near the level of the stream, and representing the dissected remnants of an abandoned flood plain, stream bed, or valley floor produced during a former state of erosion or deposition. Erosional surfaces cut into bedrock and thinly mantled with stream deposits (**alluvium**) are designated "**strath terraces**." Remnants of constructional valley floors thickly mantled with **alluvium** are termed alluvial terraces.

strip cropping–See **tillage**, *strip cropping*.

strip planting–See **tillage**, *strip planting*.

strip till planting–See **tillage**, *strip till planting*.

structural charge–The charge (usually negative) on a mineral resulting from isomorphous substitution within the mineral layer. (Expressed as moles (mol) or centimoles (cmol) of charge per kilogram of clay.)

structure–See **soil structure** or **crystal structure**.

stubble mulch–See **tillage**, *stubble mulch.*

Subarctic Brown Forest soils–Soils similar to Brown Forest soils except having more shallow sola and average temperatures of <5°C at 18 inches or more below the surface. (Not used in current U.S. system of soil taxonomy.)

subbing–See **irrigation**, *subbing.*

subsoiling–See **tillage**, *subsoiling.*

substrate–(i) That which is laid or spread under an underlying layer, such as the subsoil. (ii) The substance, base, or nutrient on which an organism grows. (iii) Compounds or substances that are acted upon by enzymes or catalysts and changed to other compounds in the chemical reaction.

substratum–Any layer lying beneath the soil solum, either conforming or unconforming.

subsurface tillage–See **tillage**, *subsurface tillage.*

sulfidic material–Waterlogged material or organic material that contains 7.5 g kg^{-1} or more of sulfide-sulfur.

sulfur cycle–The sequence of transformations undergone by sulfur wherein it is used by living organisms, transformed upon death and decomposition of the organism, and ultimately converted to its original oxidation state.

sulfuric horizon–A horizon composed either of mineral or organic soil material that has both pH <3.5 and **jarosite** mottles.

summation curve, particle size–A curve showing the accumulative percentage by mass of particles within increasing (or decreasing) size limits as a function of diameter; the percent by mass of each size fraction is plotted accumulatively on the ordinate as a function of the total range of diameters represented in the sample plotted on the abscissa.

summer fallow–See **tillage**, *summer fallow.*

summit–The highest point of any landform remanant, hill, or mountain.

superphosphate–A product obtained when phosphate rock is treated with H_2SO_4, H_3PO_4, or a mixture of those acids.

ammoniated–A product obtained when superphosphate is treated with NH_3 or with solutions containing NH_3 and/or other NH_4–N containing compounds.

concentrated–Also called *triple* or *treble* superphosphate, made with phosphoric acid and usually containing 19 to 21% P (44 to 48% P_2O_5).

enriched–Superphosphate made with a mixture of sulfuric acid and phosphoric acid. This includes any grade between 10 and 19% P (22% and 44% P_2O_5), commonly 11 to 13% P (25 to 30% P_2O_5).

normal–Also called *ordinary* or *single* superphosphate. Superphosphate made by reaction of phosphate rock with sulfuric acid, usually containing 7 to 10% P (16 to 22% P_2O_5).

ordinary–See **superphosphate, normal**.

single–See **superphosphate, normal**.

superphosphoric acid–The acid form of polyphosphates, consisting of a mixture of orthophosphoric and polyphosphoric acids. Species distribution varies with concentration, which is typically 30 to 36% P (68 to 83% P_2O_5).

supraglacial–Carried upon, deposited from, or pertaining to the top surface of a **glacier** or ice sheet; said of meltwater streams, **till**, **drift**, etc.

surface area–The area of the solid particles in a given quantity of soil or porous medium. (i) BET surface area is that area on which gas molecules, such as N_2 or O_2, can adsorb which normally does not include the planar surface of expanding clays such as **smectites**. (ii) EGME surface area is that area on which ethylene glycol monoethyl ether can adsorb which normally includes the planar surface of expanding clays such as **smectites**. See also **specific surface**.

surface creep–See **erosion**, *surface creep*.

surface runoff–See **runoff**.

surface sealing–The deposition by water, orientation and/or packing of a thin layer of fine soil particles on the immediate surface of the soil, greatly reducing its water permeability.

surface soil–The uppermost part of the soil, ordinarily moved in **tillage**, or its equivalent in uncultivated soils and ranging in depth from 7 to 25 cm. Frequently designated as the **plow layer**, the *surface layer*, the *Ap layer*, or the *Ap horizon*. See also **topsoil**.

surface-charge density–The excess of negative or positive charge per unit of surface area of soil or soil mineral.

surfactant–A substance that lowers the surface tension of a liquid.

surge irrigation–See **irrigation.**

suspension–See **erosion**, *suspension*.

sustainability–Managing soil and crop cultural practices so as not to degrade or impair environmental quality on or off site, and without eventually reducing yield potential as a result of the chosen practice through exhaustion of either on-site resources or non-renewable inputs.

swamp–An area saturated with water throughout much of the year but with the surface of the soil usually not deeply submerged. Usually characterized by tree or shrub vegetation. See also **marsh** and **miscellaneous areas**.

sweep–See **tillage**, *sweep*.

symbiosis–The obligatory cohabitation of two dissimilar organisms in intimate association. Often, but not always, mutually beneficial.

symmetry concentration–(no longer used in SSSA publications) That quantity of cations (or anions) equivalent to the exchange capacity of a soil. For example, if the cation exchange capacity of a soil is 10 $cmol_c$ kg^{-1} of soil, then 1 symmetry concentration is 10 cmol of any monovalent cation or 5 cmol of any divalent cation.

symmetry value–(no longer used in SSSA publications) The quantity of adsorbed ion released when one symmetry concentration of another ion is added.

synergism–(i) The nonobligatory association between organisms that is mutually beneficial. Both populations can survive in their natural environment on their own although, when formed, the association offers mutual advantages. (ii) The simultaneous actions of two or more factors that have a greater total effect together than the sum of their individual effects.

T

tactoid–The colloidal sized aggregates of phyllosilicate clay particles that can form under certain conditions of exchangeable cations and ionic strength.

tailwater–See **irrigation**, *tailwater*.

tailwater recovery–See **irrigation**, *tailwater recovery*.

talc–$Si_4Mg_3O_{10}(OH)_2$ A trioctahedral magnesium silicate mineral with a 2:1 type layer structure but without **isomorphous substitution**. May occur in soils as an inherited mineral. See also **Appendix I, Table A3**.

talud–A short, steep slope formed gradually at the downslope margin of a field by deposition against a hedge, a stone wall, or other similar barrier.

talus–Rock fragments of any size or shape (usually coarse and angular) derived from and lying at the base of a cliff or very steep rock slope. The accumulated mass of such loose, broken rock formed chiefly by falling, rolling, or sliding.

taxadjunct–A soil that is correlated as a recognized, existing soil series for the purpose of expediency. They are so like the soils of the defined series in morphology, composition, and behavior that little or nothing is gained by adding a new series.

taxon–In the context of soil survey, a class at any categorical level in the U.S. system of soil taxonomy.

taxonomic unit–See **taxon**.

TDR–See **time-domain reflectometry.**

tensile strength–The load per unit area at which an unconfined cylindrical specimen will fail in a simple tension test.

tensiometer–A device for measuring the soil-water matric potential in situ; a porous, permeable ceramic cup connected through a water-filled tube to a manometer, vacuum gauge, pressure transducer, or other pressure measuring device.

tephra–A collective term for all **clastic** volcanic materials that are ejected from a vent during an eruption and transported through the air, including **ash (volcanic)**, blocks (volcanic), **cinders, lapilli, scoria**, and pumice. **Tephra** is a general term which, unlike many volcaniclastic terms, does not denote properties of composition, visicularity, or grain size.

terminal moraine–An **end moraine** that marks the farthest advance of a **glacier** and usually has the form of a massive arcuate or concentric ridge, or complex of ridges, underlain by **till** and other **drift** types. See also **end moraine, recessional moraine, ground moraine.**

terrace–(i) A step-like surface, bordering a stream or shoreline, that represents the former position of a flood plain, lake, or sea shore. (ii) A raised, generally horizontal strip of earth and/or rock constructed along a hill on or nearly on a contour to make land suitable for tillage and to prevent accelerated erosion. (iii) An earth embankment constructed across a slope for conducting water from above at a regulated flow to prevent accelerated erosion and to conserve water.

textural classification–See **soil texture**.

texture–See **soil texture**.

thermal analysis–Measurement of changes in physical or chemical properties of materials as a function of temperature, usually heating or cooling at a uniform rate. (i) Differential thermal analysis (DTA), measures temperature difference (ΔT) between a sample and reference material. (ii) Differential scanning calorimetry (DSC), measures the differential heat flow between a sample and reference material. (iii) Thermogravimetric analysis (TGA), measures weight loss or gain.

thermal band–A general term for middle-infrared wavelengths which are transmitted through the atmosphere window at 8 to 13 μm. Occasionally also used for the windows around 3 to 6 μm.

thermal properties–Properties of a medium (soil) relative to heat content and heat transfer, such as thermal conductivity, specific heat capacity, and thermal diffusivity.

thermic–A soil temperature regime that has mean annual soil temperatures of 15°C or more but <22°C, and >5°C difference between mean summer and mean winter soil temperatures at 50 cm below the surface. Isothermic is the same except the summer and winter temperatures differ by <5°C.

thermogenic soils–Soils with properties that have been influenced primarily by high temperature as the dominant soil-formation factor; developed in subtropical and equatorial regions.

thermophile–See **thermophilic organism**.

thermophilic organisms–An organism whose optimum temperature for growth is above 45°C.

thermosequence–A group of related soils that differ, one from the other, primarily as a result of differences in temperature as a soil-formation factor.

threshold moisture content–(biological) The minimum moisture condition, measured either in terms of moisture content or moisture stress, at which biological activity just becomes measurable.

throughfall–That portion of precipitation that falls through or drips off of a plant canopy.

throw–See **tillage**, *throw*.

tidal flats–Areas of nearly flat, barren mud periodically covered by tidal waters. Normally these materials have an excess of soluble salt. A **miscellaneous area.**

tie-ridging–See **tillage**, *tie-ridging*.

tile drain–Concrete, ceramic, plastic etc. pipe, or related structure, placed at suitable depths and spacings in the soil or subsoil to enhance and/or accelerate drainage of water from the soil profile.

till–(i) Unsorted and unstratified earth material, deposited by **glacial** ice, which consists of a mixture of clay, silt, sand, gravel, stones, and boulders in any proportion. (ii) To prepare the soil for seeding; to seed or cultivate the soil.

till plain–An extensive flat to undulating surface underlain by **till.**

tillability–See **tillage**, *tillability*.

tillage–The mechanical manipulation of the soil profile for any purpose; but in agriculture it is usually restricted to modifying soil conditions and/or managing crop residues and/or weeds and/or incorporating chemicals for crop production.

  anchor–Partially burying foreign materials such as plant residues or paper mulches.

backfurrow–The resulting ridge of soil turned up when the first furrow slice is lapped over the previous soil surface when starting the plowing operation.

bed planting–A method of planting in which the seed is planted on slightly raised areas between furrows with two or more seed rows sometimes planted on each bed. See also **tillage**, *ridge planting.*

bed shaper–A soil-handling implement which forms uniform ridges of soil to predetermined shapes.

bedding–The process of preparing a series of parallel ridges, usually no wider than two crop rows, separated by shallow furrows. The resulting structures are beds.

block (thinning, checking)–To remove plants from a row with hoes or other cutting devices as a means of reducing and uniformly spacing plants.

broadcast planting–A uniform planting of seeds distributed over the entire planted area.

broadcast tillage (total surface tillage, full-width tillage)–Manipulation of the entire surface area by tillage implements as contrasted to partial manipulation in bands or strips.

burying–Covering foreign materials or bodies intact, such as drain liners, tile lines, communication wires, or plant residues.

chemical fallow (eco-fallow)–A special case of fallowing in which all vegetative growth is killed or prevented by use of chemicals; tillage for other purposes may or may not be used.

chisel–To break up soil using closely spaced gangs of narrow shank-mounted tools. It may be performed at other than the normal plowing depth. Chiseling at depths >40 cm is usually termed subsoiling.

clean tillage (clean culture, clean cultivation)–A process of plowing and cultivation which incorporates all residues and prevents growth of all vegetation except the particular crop desired during the growing season.

combined tillage operations–The simultaneous operation of two or more different types of tillage tools (on the same implement frame) or implements (subsoiler-lister, lister planter, or plow planter) to simplify control or reduce the number of trips over the field.

conservation tillage–Any tillage sequence, the object of which is to minimize or reduce loss of soil and water; operationally, a tillage or tillage and planting combination which leaves a 30% or greater cover of crop residue on the surface.

contour tillage–Performing the tillage operations and planting on the contour within a given tolerance.

controlled traffic–A farming system, including tillage in which the wheel tracks of all operations are confined to fixed paths so that recompaction of soil by traffic (traction or transport) does not occur outside the selected paths.

conventional tillage–Primary and secondary tillage operations normally performed in preparing a seedbed and/or cultivating for a given crop grown in a given geographical area, usually resulting in <30% cover of crop residues remaining on the surface after completion of the tillage sequence.

crop residue management–Disposition of stubble, stalks, and other crop residues by tillage operations. (i) To remove residues from the soil surface (burying); (ii) To anchor residues partially in the surface soil while leaving the residues partially exposed at the surface (mulch tillage); (iii) To leave residues entirely at the soil surface intact or cut into smaller pieces. (Residues may be removed by nontillage methods, i.e., harvesting, burning, grazing, etc.)

crop residue management system–The operation and management of crop land to maintain stubble, stalks, and other crop residue on the surface to prevent wind and water erosion, to conserve water, and to decrease evaporation.

cross cultivation–The tillage of a field, orchard, etc., in which the field is cultivated in one direction followed by cultivation at some angle between 10 and 90° from the preceding tillage.

crushing–Applying forces to the soil surface to destroy the integrity of aggregates or clods.

cultipack–A broadcast soil crushing and firming operation utilizing wide rollers having corrugated or jagged working surfaces.

cultivation–Shallow tillage operations performed to create soil conditions conducive to improved aeration, infiltration, and water conservation, or to control weeds.

cultivation (weeding)–Tillage action which lightly tills the surface 1-2 cm of soil for the purpose of destroying weeds.

cutting–Severing soil by a slicing action that minimizes any other type of failure, such as shear.

dam (pitting, basin listing)–Forming pits, small basins, or waterholding cavities at intervals with appropriate equipment.

dammer-diker–See **tillage**, *reservoir tillage*.

deadfurrow–The furrow resulting where land plowed in one direction abuts with land plowed in the opposite direction, i.e. at the completion of each plowed section of a field.

dig–To breakup, invert, or remove the soil with a spade, plow, or other implement; or to bring to the surface (as in harvesting potatoes or disturbing subterranean root and stem structures of weeds) with mechanical tools.

drag–To draw planks or other heavy, rigid implements with wide surfaces across the soil surface to crush clods and level or smooth the surface.

eco-fallow or *ecofallow*–See **tillage**, *chemical fallow*.

fallow–The practice of leaving land either uncropped and weed-free, or with volunteer vegetation during at least one period when a crop would normally be grown; objective may be to control weeds, accumulate water, and/or available plant nutrients.

firming–A process of achieving a desirable degree of compaction.

flat planting–A planting method in which the seed is planted on flat ground without intentional surface depressions.

furrow–An opening left in the soil after a plow or disk has opened a shallow channel at the soil surface. A shallow channel cut in the soil surface, usually between planted rows, for controlling surface water and soil loss, or for conveying irrigation water.

guess row–The rows or inter row space of adjoining multiple-row equipment passes, where, due to reliance on markers for approximate positioning and guidance of tractor traffic, the inter row space will vary as the driver deviates from a perfect pattern.

harrowing–A secondary broadcast tillage operation which pulverizes, smoothes, and firms the soil in seedbed preparation, controls weeds, or incorporates material spread on the surface.

hill–To place soil up to and around crops, usually planted in rows.

hoe–To dig, scrape, or the like, with a hoe; also to control weeds or to loosen or rearrange the soil.

incorporation–Mixing of materials found on or spread upon the soil surface (e.g. fertilizers, pesticides, or crop residues) into the soil volume via tillage.

in-row subsoiling–Use of subsoiling in conjunction with traffic control or where the subsoiler tool is an integral part of the planter implement, for the purpose of having zones of maximum soil shattering located directly beneath the planted row in order to maximize root exploration or penetration of a restrictive zone shattered by the subsoiling operation.

inversion–Reversal of vertical order of occurrence of layers of soil.

landforming–Tillage operations which move soil to create desired soil configurations. Forming may be done on a large scale such as gully filling or terracing, or on a small scale such as contouring, ridging, or bedding.

land planing–A tillage operation which redistributes small quantities of soil across the soil surface to provide a more nearly level or uniformly sloped surface.

lift–To separate roots or other crop parts from soil and elevate them to the soil surface or above.

lister planting–A method of planting in which the seed is planted in the bottom of lister furrows, usually simultaneously with the opening of these furrows.

listing (middlebreaking)–A tillage and land-forming operation using a tool which turns two furrows laterally in opposite directions, thereby producing beds or ridges.

loosening Decreasing soil bulk density and increasing porosity due to the application of mechanical forces to the soil via tillage.

minimum tillage–The minimum use of primary and/or secondary tillage necessary for meeting crop production requirements under the existing soil and climatic conditions, usually resulting in fewer tillage operations than for conventional tillage.

mixing–Blending of soil layers into the soil mass.

moldboard plowing–See **tillage**, *plowing*.

mulch–(i) Any material such as straw, sawdust, leaves, plastic film, loose soil, etc., that is spread or formed upon the surface of the soil to protect the soil and/or plant roots from the effects of raindrops, soil crusting, freezing, evaporation, etc. (ii) To apply mulch to the soil surface.

mulch farming–A system of tillage and planting operations which maintains a substantial amount of plant residues or other mulch on the soil surface.

mulch tillage–Tillage or preparation of the soil in such a way that plant residues or other materials are left to cover the surface; also, mulch farming, trash farming, stubble mulch tillage, plowless farming; operationally, a full-width tillage or tillage and planting combination that leaves >30% of the surface covered with crop residue.

narrow row planting–A method of planting in which the seed is planted in uncommonly narrow rows for the given crop to hasten canopy coverage and reduce cultivation requirement.

non-inversive tillage–Tillage which does not mix (or minimizes the mixing of) soil horizons or does not vertically mix soil within a horizon.

no-tillage (zero tillage) system–A procedure whereby a crop is planted directly into the soil with no primary or secondary tillage since harvest of the previous crop; usually a special planter is necessary to prepare a narrow, shallow seedbed immediately surrounding the seed being planted. No-till is sometimes practiced in combination with subsoiling to facilitate seeding and early root growth, whereby the surface residue is left virtually undisturbed except for a small slot in the path of the subsoil shank.

once-over tillage–A system whereby all tillage preparatory for planting is done in one operation or trip over the field.

oriented tillage–Tillage operations which bear specific relations in direction with respect to the sun, prevailing winds, previous tillage operations, or field base lines.

pans–Horizons or layers in soils which are highly compacted, indurated, or very high in clay content relative to the layer immediately above.

paraplow–A type of non-inversive subsoiling implement, designed to enhance lateral direction of shattering force using broad, angled subsoil lifting surfaces.

paratill–A variation on the mounting of paraplow subsoiling implements to allow greater ease of use in row crops, and/or to leave specific non-shattered zones between rows to provide traction and support for vehicle or tractor traffic.

plowing–A primary broadcast tillage operation which is performed to shatter soil with partial to complete inversion, usually to depths greater than 20 cm.

plow layer–The greatest depth of soil exhibiting mixing or inversion by surface tillage operations.

plowless farming–Tilling soil without moldboard plowing so that inversion and/or residue burying is intentionally reduced.

plow pan–A pan created by plowing at the depth of tillage, largely the result of the common practice of dropping the tractor wheels of one side of the tractor into the dead furrow for steering while performing the plowing operation.

plow-planting–The plowing and planting of land in a single trip over the field by drawing both plowing and planting tools with the same power sources.

pre-(post) emergence tillage–Tillage operations which occur before (after) crop emergence.

pre-(post) harvest tillage–Tillage operations which occur before (after) crop harvest.

pre-(post) planting tillage–Tillage operations which occur before (after) the crop is planted.

pressure pan (traffic sole, hard pan, plow pan, tillage pan, traffic pan, plow sole, compacted layer)–An induced subsurface soil horizon or layer having a higher bulk density and lower total porosity than the soil material directly above and below, but similar in particle size analysis and chemical properties. The pan is usually found just below the maximum depth of primary tillage and frequently restricts root development and water movement.

primary tillage–Tillage at any time which constitutes the initial, major soil manipulation operation. It is normally a broadcast operation designed to loosen the soil or reduce soil strength, anchor or bury plant materials and fertilizers, and rearrange aggregates.

reduced tillage–A tillage system in which the total number of tillage operations preparatory for seed planting is reduced from that normally used on that particular field or soil. See also **tillage**, *minimum tillage*.

reservoir tillage (damming, pitting, basin listing, furrow diking, dammer diking)–Forming pits, small basins, or water-holding cavities at intervals with a furrow diker or other appropriate equipment.

residue processing–Operations which cut, crush, shred, or otherwise break (fracture) residues in a step preparatory to tillage, harvesting, or planting operations.

ridge–To form a raised longitudinal mound of soil by a lister or other tillage tool.

ridge planting–A method of planting crops on ridges formed through tillage operations. Usually only one seed row is planted on each ridge.

ridge tillage–A tillage system in which ridges are reformed atop the planted row by cultivation, and the ensuing row crop is planted into ridges formed the previous growing season. See also **tillage**, *ridge planting*.

rod weeding–Control or eradication of weeds and soil firming by means of pulling a longitudinally rotating rod below the soil surface. The rod rotates about an axis perpendicular to the line of travel, and pulls or cuts off weeds with minimum disturbance of trash on or near the ground surface.

rolling–A broadcast, secondary tillage operation which crushes clods and compacts or firms and smoothes the soil by the action of ground-driven, rotating cylinders. See also **tillage**, *cultipack*.

root bed–The soil profile modified by tillage or amendments for more effective use by plant roots.

rotary hoeing–A shallow tillage operation employing ground-driven rotary motion of the tillage tool to shatter and mix soil and control small weed seedlings.

rotary tilling–A tillage operation employing power driven rotary motion of the tillage tool to loosen, shatter, and mix soil.

scarifying–To loosen the topsoil aggregates by means of raking the soil surface with a set of sharp teeth.

secondary tillage–Any of a group of separate or distinct tillage operations, following primary tillage, that is designed to provide specific soil conditions for any reason, such as seeding.

seedbed–The tillage manipulated soil layer which affects the germination and emergence of crop seeds.

shatter–General fragmentation of a rigid or brittle soil mass.

shearing–Separating parts of a soil mass by applying shearing stresses.

slit tillage–Use of narrow straight coulters or knives to open slices of 5-10 mm in width in soil which penetrate to beneath a shallow root restrictive layer, allowing precision planted seeds to develop root systems which penetrate the restrictive layer, without requiring large-scale profile disruption or shattering, and the horsepower or energy needed to accomplish such operations.

slit planting (slot planting)–A method of planting crops that involves no seedbed preparation other than opening a fine slit in the soil (usually with a coulter attached to the planter) to place the seed at some intended depth. Herbicides are usually sprayed shortly before, at, or after planting when performed in reduced tillage systems.

sod planting–A method of planting in sod with little or no tillage.

soil management–The combination of all tillage operations, cropping practices, fertilizer, lime, and other treatments conducted on or applied to the soil for the production of plants.

strip cropping (field strip cropping, contour strip cropping)–The practice of growing two or more crops in alternating strips along contours, often perpendicular to the prevailing direction of wind or surface water flow.

strip planting (strip till planting)–A method of simultaneous tillage and planting in isolated bands of varying width, separated by bands of erect residues essentially undisturbed by tillage.

strip tillage (partial-width tillage)–Tillage operations performed in isolated bands separated by bands of soil essentially undisturbed by the particular tillage equipment.

strip-till planting–An area 30 to 50 cm wide is tilled sufficiently through living mulch or standing residue to form a seedbed for each row. At planting or at first cultivation, the remaining mulch in the row middle is cut loose, killed or retarded.

stubble mulch–The stubble of crops or crop residues left essentially in place on the land as a surface cover before and during the preparation of the seedbed and at least partly during the growing of a succeeding crop.

stubble mulch tillage–See **tillage**, *mulch tillage*; **tillage**, *plowless farming.*

subsoiling–Any treatment to non-inversively loosen soil below the Ap horizon with a minimum of vertical mixing of the soil. Any treatment to fracture and/or shatter soil with narrow tools below the depth of normal tillage without inversion and with a minimum mixing of the soil. This loosening is usually performed by lifting action or other displacement of soil dry enough so that shattering occurs.

subsurface tillage–Tillage which confines most of its action (usually only fracturing and shattering) to depths below the normal depth of disc cultivation.

summer fallow–The prevention of all vegetative growth by shallow tillage in conjunction with or without herbicides during the summer months, in place of growing a crop, in order to store water for use by the next crop, or to control weed infestations.

surface tillage–Cultivating or mixing the soil to a shallow depth.

sweep–(i) Tillage with a shallow knife, blade, or sweep cultivating tool which is drawn slightly beneath the soil surface cutting plant roots and loosening the soil without inverting it, resulting in minimum incorporation of residues into the soil. (ii) A type of cultivator shovel which is wing-shaped.

throw–Aerial movement of soil in any direction resulting from momentum imparted to the soil.

tie-ridging–Joining together of ridges at certain intervals by a cross ridge to form small basins.

tillability–The degree of ease with which a soil may be manipulated for a specific purpose.

tillage action–The specific form or forms of soil manipulation performed by the application of mechanical forces to the soil with a tillage tool, such as cutting, shattering, inversion, or mixing.

tillage, deep–A primary tillage operation which manipulates soil to a greater depth than normal plowing. It may be accomplished with a large heavy-duty moldboard or disk plow which inverts the soil, or with a heavy-duty chisel plow which shatters soil. See also **tillage**, *subsoiling.*

tillage equipment (tools)–Field tools and machinery which are designed to lift, invert, stir, or pack soil, reduce the size of clods or uproot weeds, i.e., plows, harrows, disks, cultivators, and rollers.

tillage operation–Act of applying one or more tillage actions in a distinct mechanical application of force to all or part of the soil mass.

tilth–The physical condition of soil as related to its ease of tillage, fitness as a seedbed, and its impedance to seedling emergence and root penetration.

trash farming–See **tillage**, *mulch tillage;* **tillage**, *no-tillage;* **tillage**, *zero tillage;* **tillage**, *minimum tillage;* **tillage**, *plowless farming.*

turnrow (turn strip, head land)–The land at the margin of a field on which the plow or other equipment may be turned. This land may or may not be planted to a crop.

vertical mulching–A subsoiling operation in which a vertical band of mulching material is placed into the vertical slit in the soil made by the soil-opening implement.

weeding–Tillage action which lightly cultivates the soil for the purpose of destroying weeds.

wheel track planting–A practice of planting in which the seed is planted in tracks formed by wheels (usually tractor wheels) rolling immediately ahead of the planter.

zero tillage–See **tillage**, *no-tillage system.*

zone subsoiling–The practice, usually only in row crops, of maximizing subsoil shattering in certain zones along the row, while specifically preventing it in trafficked interrows, thereby maximizing crop response without impairing traction of vehicles or tractors later entering the field. Can be accomplished with in-row subsoilers, but usually seeks a larger shattering zone, such as the type obtained with the paratill.

zone tillage–Tillage operations which differentially affect parallel zones traversed by the tillage implement machine.

tillage erosion–See **erosion**, *tillage erosion.*

tilth–See **tillage**, *tilth.*

time-domain reflectometry–A method that uses the timing of wave reflections to determine the properties of various materials, such as the dielectric constant of soil as an indication of water content.

todorokite–$(Na, Ca, K, Ba, Mn^{2+})_2 Mn_4O_{12}\cdot 3H_2O$ A black manganese oxide that occurs in soils and in the weathered regolith of sediments. It has a tunnel structure.

toeslope–The hillslope position that forms a gently inclined surface at the base of a slope. Toeslopes in profile are commonly gentle and linear, and are constructional surfaces forming the lower part of a slope continuum that grades to a valley or closed depression.

top dressing–An application of fertilizer to a soil surface, without incorporation, after the crop is established.

toposequence–A sequence of related soils that differ, one from the other, primarily because of topography as a soil-formation factor.

topsoil–(i) The layer of soil moved in cultivation. Frequently designated as the *Ap layer* or *Ap horizon*. See also **surface soil**. (ii) Presumably fertile soil material used to topdress roadbanks, gardens, and lawns.

Torrands–**Andisols** that have an **aridic** soil moisture regime. (A suborder in the U.S. system of soil taxonomy.)

Torrerts–**Vertisols** of arid regions that if not irrigated during the year have cracks in 6 or more out of 10 years that remain closed for less than 60 consecutive days during a period when the soil temperature at a depth of 50 cm from the surface is higher than 8° C. (A suborder in the U.S. system of soil taxonomy.)

torric–A soil moisture regime defined like aridic moisture regime but used in a different category of the U.S. soil taxonomy.

Torrox–**Oxisols** that have a **torric** soil moisture regime. (A suborder in the U.S. system of soil taxonomy.)

tortuosity–The nonstraight nature of soil pores.

tortuosity factor–The reciprocal of the increase in diffusion path that an ion must take in diffusing through the water present in the soil when it moves along a concentration gradient as compared to the path in water.

total potential (of soil water)–See **soil water** and **Table 5**.

total pressure–See **soil water** and **Table 5**.

toxicity–Quality, state, or degree of the harmful effect from alteration of an environmental factor.

trace elements (i) It is no longer used in SSSA publications in reference to plant nutrition. See also **micronutrient**. (ii) In environmental applications it is those elements exclusive of the eight abundant rock-forming elements: oxygen, aluminum, silicon, iron, calcium, sodium, potassium, and magnesium.

traffic pan–See **pan, pressure or induced**.

transitional soil (intergrades)–A soil that possesses properties and distinguishing characteristics of two or more separate soils.

trash farming–See **tillage**, *mulch tillage; no-tillage; zero tillage; minimum tillage; plowless farming*.

tree-tip mound–The small mound of debris sloughed from the root plate (ball) of a tipped-over tree. Local soil horizons are commonly obliterated, which results in a heterogeneous mass of soil material.

tree-tip pit–The small pit or depression resulting from the area vacated by the root plate(ball) from tree-tip. Such pits are commonly adjacent to small mounds composed of the displaced material. Subsequent infilling usually produces a heterogeneous soil matrix.

triaxial shear test–A test in which a cylindrical specimen of soil **umbric epipedon** encased in an impervious membrane is subjected to a confining pressure and then loaded.

trickle irrigation–See **irrigation,** *trickle.*

trioctahedral–An octahedral sheet or a mineral containing such a sheet that has all of the sites filled, usually by divalent ions such as magnesium or ferrous iron. See also **phyllosilicate mineral terminology** and **dioctahedral**

Tropepts–**Inceptisols** that have a mean annual soil temperature of 8°C or more, and <5°C difference between mean summer and mean winter temperatures at a depth of 50 cm below the surface. **Tropepts** may have an **ochric epipedon** and a **cambic horizon**, or an **umbric epipedon**, or a **mollic epipedon** under certain conditions but no **plaggen epipedon**, and are not saturated with water for periods long enough to limit their use for most crops. (A suborder in the U.S. system of soil taxonomy.)

truncated–Having lost all or part of the upper soil horizon or horizons by soil removal (erosion, excavation, etc.).

tuff–A compacted deposit that is 50 percent or more volcanic ash and dust.

Tundra soils–(i) Soils characteristic of tundra regions. (ii) A **zonal** great soil group consisting of soils with dark-brown peaty layers over grayish horizons mottled with rust and having continually frozen substrata; formed under frigid, humid climates, with poor drainage, and native vegetation of lichens, moss, flowering plants, and shrubs. (Not used in current U.S. system of soil taxonomy.)

turn strip–See **tillage,** *turnrow.*

turnrow–See **tillage,** *turnrow.*

U

Udalfs–**Alfisols** that have a **udic** soil moisture regime and **mesic** or warmer soil temperature regimes. **Udalfs** generally have brownish colors throughout, and are not saturated with water for periods long enough to limit their use for most crops. (A suborder in the U.S. system of soil taxonomy.)

Udands–**Andisols** that have a **udic** soil moisture regime. (A suborder in the U.S. system of soil taxonomy.)

Uderts–**Vertisols** of relatively humid regions that have wide, deep cracks that usually remain open continuously for <60 days or intermittently for periods that total <90 days. (A suborder in the U.S. system of soil taxonomy.)

udic–A soil moisture regime that is neither dry for as long as 90 cumulative days nor for as long as 60 consecutive days in the 90 days following the summer solstice at periods when the soil temperature at 50 cm below the surface is above 5°C.

Udolls–**Mollisols** that have a **udic** soil moisture regime with mean annual soil temperatures of 8°C or more. **Udolls** have no **calcic** or **gypsic horizon**, and are not saturated with water for periods long enough to limit their use for most crops. (A suborder in the U.S. system of soil taxonomy.)

Udox–**Oxisols** that have a **udic** soil moisture regime. (A suborder in the U.S. system of soil taxonomy.)

Udults–Ultisols that have low or moderate amounts of organic carbon, reddish or yellowish **argillic horizons**, and a **udic** soil moisture regime. **Udults** are not saturated with water for periods long enough to limit their use for most crops. (A suborder in the U.S. system of soil taxonomy.)

Ultisols–Mineral soils that have an **argillic horizon** with a base saturation of <35% when measured at pH 8.2. **Ultisols** have a mean annual soil temperature of 8°C or higher. (An order in the U.S. system of soil taxonomy.)

Umbrepts–Inceptisols formed in cold or temperate climates that commonly have an **umbric epipedon**, but they may have a **mollic** or an **anthropic epipedon** 25 cm or more thick under certain conditions. These soils are not dominated by **amorphous materials** and are not saturated with water for periods long enough to limit their use for most crops. (A suborder in the U.S. system of soil taxonomy.)

umbric epipedon–A surface layer of mineral soil that has the same requirements as the **mollic epipedon** with respect to color, thickness, organic carbon content, consistence, structure, and phosphorus content, but that has a base saturation <50% when measured at pH 7.

unaccommodated–Applied to peds. Virtually none of the faces of adjoining peds are molds of each other.

unconformity–A substantial break or gap in the geologic record where a unit is overlain by another that is not in stratigraphic succession.

underfit stream–A stream that appears to be too small to have eroded the valley in which it flows; a stream whose volume is greatly reduced or whose meanders show a pronounced shrinkage in radius. It is a common result of drainage changes effected by capture, glaciers, or climatic variations.

underground runoff (seepage)–Water that seeps toward stream channels after infiltration into the ground.

undifferentiated group–A kind of map unit used in soil surveys comprised of two or more taxa components that are not consistently associated geographically. Delineations show the size, shape, and location of a *landscape unit* composed of one or the others, or all of two or more component soils that have the same or very similar use and management for specified common uses. Inclusions may occur up to some allowable limit. See also **component soil**, **soil consociation**, **soil complex**, **soil association**, **miscellaneous areas**.

unit structure–See **phyllosilicate mineral terminology**.

Universal Soil Loss Equation (USLE)–See **erosion**, *universal soil loss equation (USLE)*.

unsaturated flow–The movement of water in soil in which the pores are not filled to capacity with water.

upper plastic limit–See **liquid limit**.

urban land–Areas so altered or obstructed by urban works or structures that identification of soils is not feasible. A also **miscellaneous area**.

Ustalfs–Alfisols that have an **ustic** soil moisture regime and **mesic** or warmer soil temperature regimes. **Ustalfs** are brownish or reddish throughout and are not saturated with water for periods long enough to limit their use for most crops. (A suborder in the U.S. system of soil taxonomy.)

Ustands–Andisols that have an **ustic** soil moisture regime. (A suborder in the U.S. system of soil taxonomy.)

Usterts–Vertisols of temperate or tropical regions that have wide, deep cracks that usually remain open for periods that total >90 days but do not remain open continuously throughout the year, and have either a mean annual soil temperature of 22°C or more or a mean summer and mean winter soil temperature at 50 cm below the surface that differ by <5°C or have cracks that open and close more than once during the year. (A suborder in the U.S. system of soil taxonomy.)

ustic–A soil moisture regime that is intermediate between the **aridic** and **udic** regimes and common in temperate subhumid or semiarid regions, or in tropical and subtropical regions with a monsoon climate. A limited amount of water is available for plants but occurs at times when the soil temperature is optimum for plant growth.

Ustolls–Mollisols that have an **ustic** soil moisture regime and **mesic** or warmer soil temperature regimes. **Ustolls** may have a **calcic, petrocalcic,** or **gypsic horizon,** and are not saturated with water for periods long enough to limit their use for most crops. (A suborder in the U.S. system of soil taxonomy.)

Ustox–Oxisols that have an **ustic** moisture regime and either **hyperthermic** or **isohyperthermic** soil temperature regimes or have <20 kg organic carbon in the surface cubic meter. (A suborder in the U.S. system of soil taxonomy.)

Ustults–Ultisols that have low or moderate amounts of organic carbon, are brownish or reddish throughout, and have an **ustic** soil moisture regime. (A suborder in the U.S. system of soil taxonomy.)

V

vadose water–Water in the **vadose zone**.

vadose zone–The aerated region of soil above the permanent water table.

value, color–The degree of lightness or darkness of a color in relation to a neutral gray scale. On a neutral gray scale, value extends from pure black to pure white; one of the three variables of color. See also **Munsell color system, hue,** and **chroma**.

vapor flow–The gaseous flow of water vapor in soils from a moist or warm zone of higher potential to a drier or colder zone of lower potential.

variable charge–A solid surface carrying a net electrical charge which may be positive, negative, or zero, depending on the activity of one or more species of a potential-determining ions in the solution phase contacting the surface. For minerals and other materials common in soils(e.g. soil **organic matter,** and oxides), the potential-determining ion usually is H^+ or OH^-, but any ion that forms a complex with the surface may be potential-determining. See also **constant-potential surface** and **pH dependent charge**.

variant–See **soil variant**.

varnish–See **desert varnish**.

varve–A sedimentary layer, lamina, or sequence of laminae, deposited in a body of still water within 1 year; specifically, a thin pair of graded glaciolacustrine layers seasonally deposited, usually by meltwater streams, in a glacial lake or other body of still water in front of a **glacier**.

vegetative cell–The growing or feeding form of a microbial cell, as opposed to a resistant resting form.

ventifact–A stone or pebble that has been shaped, worn, faceted, or polished by the abrasive action of windblown sand, usually under arid conditions. When the pebble is at the ground surface, as in a desert pavement, the upper part is polished while the lower or below ground part is angular or subangular.

vermiculite–A highly charged, averages about 159 $cmol_c$ kg^{-1} for soil vermiculites but has a very wide range, layer silicate of the 2:1 type that is formed from mica. It is characterized by adsorption preference for potassium, ammonium, and cesium over smaller exchange cations. It may be di- or trioctahedral. See also **Appendix I, Table A3**.

vertical mulching–See **tillage**, *vertical mulching*.

Vertisols–Mineral soils that have 30% or more clay, deep wide cracks when dry, and either **gilgai** microrelief, intersecting **slickensides**, or wedge-shaped structural aggregates tilted at an angle from the horizon. (An order in the U.S. system of soil taxonomy.)

very coarse sand–A soil separate. See also **soil separates**.

very fine sand–(i) A soil separate. See also **soil separates**. (ii) A soil textural class. See also **soil texture**.

very fine sandy loam–A soil textural class. See also **soil texture**.

vesicles–(i) Unconnected voids with smooth walls. (ii) Spherical structures, formed intracellularly, by vesicular arbuscular **endomycorrhizal** fungi.

vesicular arbuscular–A common **endomycorrhizal** association produced by phycomycetous fungi of the family *Endogonaceae*. Host range includes most agricultural and horticultural crops.

Vitrands–**Andisols** that have 1500-kPa water retention of <15 % on air dry <30 % on undried samples throughout 60 % of the thickness either; a) within 60 cm of the soil surface or top of an organic layer with **andic** properties, whichever is shallower if there is no **lithic, paralithic contact, duripan,** or **petrocalcic horizon** within that depth, or b) between the mineral soil surface or top of an organic layer with **andic** properties, whichever is shallower and a **lithic, paralithic contact, duripan,** or **petrocalcic horizon**. (A suborder in the U.S. system of soil taxonomy).

void ratio–The ratio of the volume of soil pore (or void) space to the solid-particle volume.

volcaniclastic–Pertaining to the entire spectrum of fragmental materials with a preponderance of clasts of volcanic origin. The term includes not only pyroclastic materials but also epiclastic deposits derived from volcanic source areas by normal processes of mass movement and stream erosion. Examples: welded tuff, volcanic **breccia**.

volume weight–(no longer used in SSSA publications) See **bulk density**.

volumetric water content–The soil-water content expressed as the volume of water per unit bulk volume of soil.

vughs–Relatively large voids, usually irregular and not normally interconnected with other voids of comparable size; at the magnifications at which they are recognized they appear as discrete entities.

W

wasteland–Land not suitable for, or capable of, producing materials or services of value. See also **miscellaneous areas**.

water (or matric) suction–(no longer used in SSSA publications) The preferred term is **matric potential**. See **soil water**, *soil water potential*.

water conductivity–See **soil water**, *hydraulic conductivity*.

water content–See **soil water**, *water content*.

water drop penetration time (WDPT)–A measure of soil water repellency which uses the imbibition time of drops of prescribed aqueous solutions as a discriminator.

water repellent soil–See **soil hydrophobicity, water drop penetration time.**

water table–The upper surface of ground water or that level in the ground where the water is at atmospheric pressure.

water table, perched–A saturated layer of soil which is separated from any underlying saturated layers by an unsaturated layer.

water tension–See **soil water**, *soil water potential*.

water use efficiency–Dry matter or harvested portion of crop produced per unit of water consumed.

waterlogged–Saturated or nearly saturated with water.

water-release curve–See **soil water characteristic**.

water-retention curve– See **soil water characteristic**.

water-soluble phosphate–That part of the phosphorus in a fertilizer that is soluble in water as determined by prescribed chemical tests.

water-stable aggregate–A soil aggregate which is stable to the action of water such as falling drops, or agitation as in wet-sieving analysis.

weathering–All physical and chemical changes produced in rocks, at or near the earth's surface, by atmospheric agents. See also **chemical** and **physical weathering**.

weeding–See **tillage**, *weeding*.

wetland–A transitional area between aquatic and terrestrial ecosystems that is inundated or saturated for long enough periods to produce **hydric soils** and support hydrophytic vegetation. See also **bay, bog, fen, marsh, pocosin, swamp**, and **tidal flats**.

wet prairies–See **marsh**.

wettability–See **soil wettability.**

wetting front–The boundary between the wetted region and the dry region of soil during infiltration.

wheel track planting–See **tillage**, *wheel track planting*.

wild-flooding–See *irrigation, wild-flooding*.

wilting coefficient–(no longer used in SSSA publications) A calculated value of the approximate wilting point or permanent wilting percentage. Calculated as follows:

Wilting coefficient = Hygroscopic coefficient/0.68

or

Wilting coefficient = Moisture equivalent/1.84

windbreak–See **erosion**, *windbreak*.

windthrow mound–See **tree-tip mound**.

X

xenobiotic–A compound foreign to biological systems. Often refers to human-made compounds that are resistant or recalcitrant to biodegradation and/or decomposition.

Xeralfs–**Alfisols** that have a **xeric** soil moisture regime. **Xeralfs** are brownish or reddish throughout. (A suborder in the U.S. system of soil taxonomy.)

Xerands–**Andisols** that have a **xeric** soil moisture regime. (A suborder in the U.S. system of soil taxonomy.)

Xererts–**Vertisols** that have a **thermic, mesic,** or **frigid** soil temperature regime and if not irrigated, cracks that remain both 5 cm or more wide through a thickness of 25 cm or more within 50 cm of the mineral soil surface for 60 or more consecutive days during 90 days following the summer solstice and closed 60 or more consecutive days during the 90 days following the winter solstice. (A suborder in the U.S. system of soil taxonomy.)

xeric–A soil moisture regime common to Mediterranean climates that have moist cool winters and warm dry summers. A limited amount of water is present but does not occur at optimum periods for plant growth. Irrigation or summer-fallow is commonly necessary for crop production.

Xerolls–**Mollisols** that have a **xeric** soil moisture regime. **Xerolls** may have a **calcic, petrocalcic,** or **gypsic horizon,** or a **duripan**. (A suborder in the U.S. system of soil taxonomy.)

Xerults–**Ultisols** that have low or moderate amounts of organic carbon, are brownish or reddish throughout, and have a **xeric** soil moisture regime. (A suborder in the U.S. system of soil taxonomy.)

Y

yield–The amount of a specified substance produced (e.g., grain, straw, total dry matter) per unit area.

yield curve–A graphical representation of nutrient application rate or availability versus crop yield or nutrient uptake.

yield goal–The yield that a producer expects to achieve, based on overall management imposed and past production records.

yield, sustained–A continual, annual, or periodic yield of plants or plant material from an area; implies management practices which will maintain the productive capacity of the land, be economically feasible, and maintain environmental integrity of the ecosystem.

Z

zero point of charge–See **point of zero net charge**.

zero tillage–See **tillage**, *no-tillage system.*

zeta potential–See **electrokinetic potential**.

zonal soil–(i) A soil characteristic of a large area or zone. (ii) One of the three primary subdivisions (orders) in soil classification as used in the USA. (Not used in current U.S. system of soil taxonomy.)

zone subsoiling–See **tillage**, *zone subsoiling.*

zone tillage–See **tillage**, *zone tillage.*

zymogenous flora–So-called opportunistic organisms found in soils in large numbers immediately following addition of a readily decomposable organic substrate. Synonymous with **copiotrophs**.

APPENDIX I — TABULAR INFORMATION

Table A1. Outline of the U.S. soil classification system (Soil taxonomy) revised 8/17/94

Orders	Suborders	Great Groups	Orders	Suborders	Great Groups
Alfisols	Aqualfs	Albaqualfs	Andisols	Aquands	Placaquands
		Duraqualfs	(cont.)	(cont.)	Vitraquands
		Endoaqualfs			
		Epiaqualfs		Cryands	Fluvicryands
		Fragiaqualfs			Gelicryands
		Glossaqualfs			Haplocryands
		Kandiaqualfs			Hydrocryands
		Natraqualfs			Melanocryands
		Plinthaqualfs			Vitricryands
		Umbraqualfs			
				Torrands	Vitritorrands
	Boralfs	Cryoboralfs			
		Eutroboralfs		Xerands	Haploxerands
		Fragiboralfs			Melanoxerands
		Glossoboralfs			Vitritorrands
		Natriboralfs			
		Paleboralfs		Vitrands	Udivitrands
					Ustivitrands
	Ustalfs	Durustalfs			
		Haplustalfs		Ustands	Durustands
		Kandiustalfs			Haplustands
		Kanhaplustalfs			
		Natrustalfs		Udands	Durudands
		Paleustalfs			Fluvudands
		Plinthustalfs			Hapludands
		Rhodustalfs			Hydrudands
					Melanudands
	Xeralfs	Durixeralfs			Placudands
		Fragixeralfs			
		Haploxeralfs	Aridisols	Cryids	Argicryids
		Natrixeralfs			Calcicryids
		Palexeralfs			Gypsicryids
		Plinthoxeralfs			Haplocryids
		Rhodoxeralfs			Petrocryids
					Salicryids
	Udalfs	Agrudalfs			
		Ferrudalfs		Salids	Aquisalids
		Fragiudalfs			Haplosalids
		Fraglossudalfs			
		Glossudalfs		Durids	Argidurids
		Hapludalfs			Haplodurids
		Kandiudalfs			Natridurids
		Kanhapludalfs			
		Natrudalfs		Gypsids	Argigypsids
		Paleudalfs			Calcigypsids
		Rhodudalfs			Haplogypsids
					Natrigypsids
Andisols	Aquands	Cryaquands			Petrogypsids
		Duraquands			
		Endoaquands			
		Epiaquands			
		Melanaquands			

Table A1. (continued)

Orders	Suborders	Great Groups	Orders	Suborders	Great Groups
Aridisol (cont.)	Argids	Calciargids	Histosols	Folists	Borofolists
		Gypsiargids			Cryofolists
		Haplargids			Medifolists
		Natrargids			Tropofolists
		Paleargids			
		Petroargids		Fibrists	Borofibrists
					Cryofibrists
	Calcids	Haplocalcids			Luvifibrists
		Petrocalcids			Medifibrists
					Sphagnofibrists
	Cambids	Anthracambids			Tropofibrists
		Aquicambids			
		Haplocambids		Hemists	Borohemists
		Petrocambids			Cryohemists
					Luvihemists
Entisols	Aquents	Cryaquents			Medihemists
		Endoaquents			Sulfihemists
		Epiaquents			Sulfohemists
		Fluvaquents			Tropohemists
		Hydraquents			
		Psammaquents		Saprists	Borosaprists
		Sulfaquents			Cryosaprists
					Medisaprists
	Arents	Torriarents			Sulfisaprists
		Udarents			Sulfosaprists
		Ustarents			Troposaprists
		Xerarents			
			Inceptisols	Aquepts	Cryaquepts
	Psamments	Cryopsamments			Endoaquepts
		Quartzipsamments			Epiaquepts
		Torrippsamments			Fragiaquepts
		Tropopsamments			Halaquepts
		Udipsamments			Humaquepts
		Ustipsamments			Placaquepts
		Xeropsamments			Plinthaquepts
					Sulfaquepts
	Fluvents	Cryofluvents			Tropaquepts
		Torrifluvents			
		Tropofluvents		Plaggepts	Plaggepts
		Udifluvents			
		Ustifluvents		Tropepts	Dystropepts
		Xerofluvents			Eutropepts
					Humitropepts
	Orthents	Cryorthents			Sombritropepts
		Torriorthents			Ustropepts
		Troporthents			
		Udorthents		Umbrepts	Cryumbrepts
		Ustorthents			Fragiumbrepts
		Xerorthents			Haplumbrepts
					Xerumbrepts

Table A1. (continued)

Orders	Suborders	Great Groups	Orders	Suborders	Great Groups
Incep- tisols	Ochrepts	Cryochrepts Durochrepts Dystrochrepts Eutrochrepts Fragiochrepts Sulfochrepts Ustochrepts Xerochrepts	Oxisols	Aquox	Acraquox Eutraquox Haplaquox Plinthaquox
				Torrox	Acrotorrox Eutrotorrox Haplotorrox
Mollisols	Albolls	Argialbolls Natralbolls		Ustox	Acrustox Eutrustox Haplustox
	Aquolls	Argiaquolls Calciaquolls Cryaquolls Duraquolls Endoaquolls Epiaquolls Natraquolls			Kandiustox Sombriustox
				Perox	Acroperox Eutroperox Haploperox Kandiperox Sombriperox
	Rendolls	Rendolls		Udox	Acrudox Eutrudox Hapludox Kandiudox Sombriudox
	Xerolls	Argixerolls Calcixerolls Durixerolls Haploxerolls Natrixerolls Palexerolls			
			Spodosols	Aquods	Alaquods Cryaquods Duraquods Endoaquods Epiaquods Fragiaquods Placaquods
	Borolls	Argiborolls Calciborolls Cryoborolls Haploborolls Natriborolls Paleborolls Vermiborolls			
				Cryods	Duricryods Haplocryods Humicryods Placocryods
	Ustolls	Argiustolls Calciustolls Durustolls Haplustolls Natrustolls Paleustolls Vermustolls		Humods	Durihumods Fragihumods Haplohumods Placohumods
	Udolls	Argiudolls Calciudolls Hapludolls Paleudolls Vermudolls		Orthods	Alorthods Durorthods Fragiorthods Haplorthods Placorthods

Table A1. (continued)

Orders	Suborders	Great Groups	Orders	Suborders	Great Groups
Ultisols	Aquults	Albaquults	Vertisols	Aquerts	Calciaquerts
		Endoaquults			Duraquerts
		Epiaquults			Dystraquerts
		Fragiaquults			Endoaquerts
		Kandiaquults			Epiaquerts
		Kanhaplaquults			Natraquerts
		Paleaquults			Salaquerts
		Plinthaquults			
		Umbraquults		Cryerts	Haplocryerts
					Humicryerts
	Humults	Haplohumults			
		Kandihumults		Xererts	Calcixererts
		Kanhaplohumults			Durixererts
		Palehumults			Haploxererts
		Plinthohumults			
		Sombrihumults		Torrerts	Calcitorrerts
					Gypsitorrerts
	Udults	Fragiudults			Haplotorrerts
		Hapludults			Salitorrerts
		Kandiudults			
		Kanhapludults		Usterts	Calciusterts
		Paleudults			Dystrusterts
		Plinthudults			Gypsiusterts
		Rhodudults			Haplusterts
					Salusterts
	Ustults	Haplustults			
		Kandiustults		Uderts	Dystruderts
		Kanhaplustults			Hapluderts
		Paleustults			
		Plinthustults			
		Rhodustults			
	Xerults	Haploxerults			
		Palexerults			

Table A2. Prefixes and their connotations for names of great groups in the U.S. soil classification system (Soil taxonomy).

Prefix	Connotation of prefix
acr	Extreme weathering
agr	An agric horizon
al	High exchangeable aluminum
alb	An albic horizon
anthr	An anthropic horizon
aqu	Evidence of wetness
arg	An argillic horizon
bor	Cool
calc	A calcic horizon
camb	A cambic horizon
cry	Cold
dur	A duripan
dystr, dys	Low base saturation
endo	Wet from below
epi	Wet, perched
eutr, eu	High base saturation
ferr	Presence of iron
fluv	Floodplain
frag	Presence of fragipan
gel	Soil temperature $<0°C$
fragloss	See the formative elements *frag* and *gloss*
gloss	Tongued
gyps	Presence of gypsic horizon
hal	Salty
hapl	Minimum horizon
hum	Presence of humus
hydr	Presence of water
kand	Presence of low activity clay
kandhapl	See kand and hapl
luv	Illuvial
med	Normal, default Histosol class
melan	Black, high C, and short-range-order minerals
nadur	See the formative elements *natr* and *dur*
natr	Presence of natric horizon
pale	Old development
petro	Hardened calcic or gypsic horizon
plac	Presence of a thin cemented layer
plag	Presence of plaggen horizon
plinth	Presence of plinthite
psamm	Sand textures
quartz	High quartz content
rhod	Dark red color
sal	Presence of salic horizon
sombr	A dark horizon
sphagn	Presence of Sphagnum moss
sulf	Sulfides or their oxidation products
torr	Torric moisture regime
trop	Continually warm and humid
ud	Udic moisture regime
umbr	Presence of umbric epipedon
ust	Ustic moisture regime
verm	Wormy, or mixed by animals
vitr	Presence of glass
xer	Xeric moisture regime

Table A3. Classification scheme for phyllosilicates related to clay minerals

Type	Group (x = charge per formula unit)	Subgroup	Species [idealized formula]†
1:1	Kaolin serpentine	Kaolins	Kaolinite $[Si_4Al_4O_{10}(OH)_8]$
			Halloysite (0.7nm) $[Si_4Al_4O_{10}(OH)_8]$ tube shape
			Halloysite (1.0nm) $[Si_4Al_4O_{10}(OH)_8 \cdot 4H_2O]$ tube shape
	$x \sim 0$	Serpentines	Chrysotile $[Si_4Mg_6O_{10}(OH)_8]$ fibrous shape, Lizardite $[Si_4Mg_6O_{10}(OH)_8]$ platy shape, Antigorite $[Si_4Mg_6O_{10}(OH)_8]$ platy or splintery shape
2:1	Pyrophyllite talc	Pyrophyllites	Pyrophyllite $[Si_4Al_2O_{10}(OH)_2]$
	$x \sim 0$	Talcs	Talc $[Si_4Mg_3O_{10}(OH)_2]$
	Smectite	Dioctahedral smectites	Montmorillonite $[Ca_{0.25}(Si_4)(Al_{1.5}Mg_{0.5})O_{10}(OH)_2]$, Beidellite $[Ca_{0.25}(Si_{3.5}Al_{0.5})(Al_2)O_{10}(OH)_2]$, Nontronite $[Ca_{0.25}(Si_{3.5}Al_{0.5})(Fe_2)O_{10}(OH)_2]$
	$x = 0.25\text{-}0.6$	Trioctahedral smectites	Saponite $[Ca_{0.34}(Si_{3.66}Al_{0.34})(Mg_3)O_{10}(OH)_2]$, Hectorite $[(Si, Al)_4(Mg, Li)_3O_{10}(OH)_2]$, Sauconite $[(Si_{3.66}Al_{0.34})(Mg, Zn)_3O_{10}(OH)_2]$
	Vermiculite	Dioctahedral vermiculites	Dioctahedral vermiculite
	$x \sim 0.6\text{-}0.9$	Trioctahedral vermiculites	Trioctahedral vermiculite
	Mica	Dioctahedral micas	Muscovite $[K(Si_3Al)(Al_2)O_{10}(OH)_2]$ Paragonite $[Na(Si_3Al)(Al_2)O_{10}(OH)_2]$
	$x \sim 1$	Trioctahedral micas	Biotite $[K(Si_3Al)(Mg, Fe^{2+})_3O_{10}(OH)_2]$ Phlogopite $[K(Si_3Al)(Mg_3)O_{10}(OH)_2]$
	Brittle mica	Dioctahedral brittle micas	Margarite $[Ca(Si_2Al_2)(Al_2)O_{10}(OH)_2]$
	$x \sim 2$	Trioctahedral brittle micas	Clintonite $[Ca(SiAl_3)(Mg_2Al)O_{10}(OH)_2]$
	Chlorite	Dioctahedral chlorites (4-5 octahedral cations per formula unit)	
	x variable	Trioctahedral chlorites (5-6 octahedral cations per formula unit)	generalized formula: $[(Si_{4-x}Al_x)^{iv}(R^{2+}, R^{3+})_3^{vi}O_{10}(OH)_2 \cdot \{(R^{2+}, R^{3+})_3^{vi}(OH)_6\}]$ Clinochore - Mg-dominant; Chamosite - Fe(II)-dominant; Pennantite - Mn^{2+}-dominant; Nimite - Ni-dominant; Baileychlore - Zn-dominant

† Only a few examples are given.

APPENDIX II–DESIGNATIONS FOR SOIL HORIZONS AND LAYERS

Three kinds of symbols are used in combination to designate horizons and layers. These are capital letters, lower case letters, and arabic numbers; capital letters are used to designate master horizons and layers; lower case letters are used as suffixes to indicate specific characteristics of the master horizon and layer; arabic numerals are used both as suffixes to indicate vertical subdivisions within a horizon or layer and as prefixes to indicate discontinuities (*Soil survey manual*, Issued October 1993. This is a revision and enlargement of USDA Handbook No. 18, the *Soil Survey Manual* issued October 1962, and supersedes it. Reference is also made to *Keys to soil taxonomy*, 6th ed. issued, 1994).

Genetic horizons are not the equivalent of the diagnostic horizons of the U.S. soil taxonomy. Designations of genetic horizons express a qualitative judgment about the vector of changes that are believed to have taken place. Diagnostic horizons are quantitatively defined features used to differentiate between taxa in U.S. system of soil taxonomy. Horizon symbols indicate the direction of presumed pedogenesis while diagnostic horizons indicate the magnitude of that expression.

Master Horizons and Layers

O horizons–Layers dominated by organic material.

A horizons–Mineral horizons that formed at the surface or below an O horizon that exhibit obliteration of all or much of the original rock structure and (i) are characterized by an accumulation of humified organic matter intimately mixed with the mineral fraction and not dominated by properties characteristic of E or B horizons; or (ii) have properties resulting from cultivation, pasturing, or similar kinds of disturbance.

E horizons–Mineral horizons in which the main feature is loss of silicate clay, iron, aluminum, or some combination of these, leaving a concentration of sand and silt particles of quartz or other resistant materials.

B horizons–Horizons that formed below an A, E, or O horizon and are dominated by obliteration of all or much of the original rock structure and show one or more of the following:

(i)   illuvial concentration of silicate clay, iron, aluminum, humus, carbonates, gypsum, or silica, alone or in combination;
(ii) evidence of removal of carbonates;
(iii) residual concentration of sesquioxides;
(iv) coatings of sesquioxides that make the horizon conspicuously lower in value, higher in chroma, or redder in hue than overlying and underlying horizons without apparent illuviation of iron;
(v)   alteration that forms silicate clay or liberates oxides or both and that forms granular, blocky, or prismatic structure if volume changes accompany changes in moisture content; or
(vi) brittleness.

C horizons or layers–Horizons or layers, excluding hard bedrock, that are little affected by pedogenic processes and lack properties of O, A, E, or B horizons. The material of C horizons may be either like or unlike that from which the solum presumably formed. The C horizon may have been modified even if there is no evidence of pedogenesis.

R layers–Hard bedrock including granite, basalt, quartzite and indurated limestone or sandstone that is sufficiently coherent to make hand digging impractical.

Transitional Horizons

Two kinds of transitional horizons are recognized. In one, the horizon is dominated by properties of one master horizon but has subordinate properties of another. Two capital latter symbols are used, such as AB, EB, BE, or BC. The master horizon symbol that is given first designates the kind of master horizon whose properties dominate the transitional horizon. In the other, distinct parts of the horizon have recognizable properties of the two kinds of master horizons indicated by the capital letters. The two capital letters are separated by a virgule (/), as E/B, B/E, or B/C. The first symbol is that of the horizon that makes up the greater volume.

AB–A horizon with characteristics of both an overlying A horizon and an underlying B horizon, but which is more like the A than the B.

EB–A horizon with characteristics of both an overlying E horizon and an underlying B horizon, but which is more like the E than the B.

BE–A horizon with characteristics of both an overlying E horizon and an underlying B horizon, but which is more like the B than the E.

BC–A horizon with characteristics of both an overlying B horizon and an underlying C horizon, but which is more like the B than the C.

CB–A horizon with characteristics of both an overlying B horizon and an underlying C horizon, but which is more like the C than the B.

E/B–A horizon comprised of individual parts of E and B horizon components in which the E component is dominant and surrounds the B materials.

B/E–A horizon comprised of individual parts of E and B horizon in which the E component surrounds the B component but the latter is dominant.

B/C–A horizon comprised of individual parts of B and C horizon in which the B horizon component is dominant and surrounds the C component.

Subordinate Distinctions Within Master Horizons and Layers

a–Highly decomposed organic material where rubbed fiber content averages <1/6 of the volume.

b–Identifiable buried genetic horizons in a mineral soil.

c–Concretions or nodules with iron, aluminum, manganese or titanium cement.

d–Physical root restriction, either natural or manmade such as dense basal till, plow pans, and mechanically compacted zones.

e–Organic material of intermediate decomposition in which rubbed fiber content is 1/6 to 2/5 of the volume.

f–Frozen soil in which the horizon or layer contains permanent ice.

g–Strong gleying in which iron has been reduced and removed during soil formation or in which iron has been preserved in a reduced state because of saturation with stagnant water.

h–Illuvial accumulation of organic matter in the form of amorphous, dispersible organic matter-sesquioxide complexes.

i–Slightly decomposed organic material in which rubbed fiber content is more than about 2/5 of the volume.

k–Accumulation of pedogenic carbonates, commonly calcium carbonate.

m–Continuous or nearly continuous cementation or induration of the soil matrix by carbonates (km), silica (qm), iron (sm), gypsum (ym), carbonates and silica (kqm), or salts more soluble than gypsum (zm).

n–Accumulation of sodium on the exchange complex sufficient to yield a morphological appearance of a natric horizon.

o–Residual accumulation of sesquioxides.

p–Plowing or other disturbance of the surface layer by cultivation, pasturing or similar uses.

q–Accumulation of secondary silica.

r–Weathered or soft bedrock including saprolite; partly consolidated soft sandstone, siltstone or shale; or dense till that roots penetrate only along joint planes and are sufficiently incoherent to permit hand digging with a spade.

s–Illuvial accumulation of sesquioxides and organic matter in the form of illuvial, amorphous, dispersible organic matter-sesquioxide complexes if *both* organic matter and sesquioxide components are significant and the value and chroma of the horizon are >3.

ss–Presence of slickensides.

t–Accumulation of silicate clay that either has formed in the horizon and is subsequently translocated or has been moved into it by illuviation.

v–Plinthite which is composed of iron-rich, humus-poor, reddish material that is firm or very firm when moist and that hardens irreversibly when exposed to the atmosphere under repeated wetting and drying.

w–Development of color or structure in a horizon but with little or no apparent illuvial accumulation of materials.

x–Fragic or fragipan characteristics that result in genetically developed firmness, brittleness, or high bulk density.

y–Accumulation of gypsum.

z–Accumulation of salts more soluble than gypsum.

Conversion Factors for SI and non-SI Units

To convert Column 1 into Column 2, multiply by	Column 1 SI Unit	Column 2 non-SI Unit	To convert Column 2 into Column 1 multiply by
Length			
0.621	kilometer, km (10^3 m)	mile, mi	1.609
1.094	meter, m	yard, yd	0.914
3.28	meter, m	foot, ft	0.304
1.0	micrometer, μm (10^{-6} m)	micron, μ	1.0
3.94×10^{-2}	millimeter, mm (10^{-3} m)	inch, in	25.4
10	nanometer, nm (10^{-9} m)	Angstrom, Å	0.1
Area			
2.47	hectare, ha	acre	0.405
247	square kilometer, km² $(10^3$ m$)^2$	acre	4.05×10^{-3}
0.386	square kilometer, km² $(10^3$ m$)^2$	square mile, mi²	2.590
2.47×10^{-4}	square meter, m²	acre	4.05×10^3
10.76	square meter, m²	square foot, ft²	9.29×10^{-2}
1.55×10^{-3}	square millimeter, mm² $(10^{-6}$ m$)^2$	square inch, in²	645
Volume			
9.73×10^{-3}	cubic meter, m³	acre-inch	102.8
35.3	cubic meter, m³	cubic foot, ft³	2.83×10^{-2}
6.10×10^4	cubic meter, m³	cubic inch, in³	1.64×10^{-5}
2.84×10^{-2}	liter, L (10^{-3} m³)	bushel, bu	35.24
1.057	liter, L (10^{-3} m³)	quart (liquid), qt	0.946
3.53×10^{-2}	liter, L (10^{-3} m³)	cubic foot, ft³	28.3
0.265	liter, L (10^{-3} m³)	gallon	3.78
33.78	liter, L (10^{-3} m³)	ounce (fluid), oz	2.96×10^{-2}
2.11	liter, L (10^{-3} m³)	pint (fluid), pt	0.473

(continued on next page)

Conversion Factors for SI and non-SI Units - (continued)

To convert Column 1 into Column 2, multiply by	Column 1 SI Unit	Column 2 non-SI Unit	To convert Column 2 into Column 1 multiply by
Mass			
2.20×10^{-3}	gram, g (10^{-3} kg)	pound, lb	454
3.52×10^{-2}	gram, g (10^{-3} kg)	ounce (avdp), oz	28.4
2.205	kilogram, kg	pound, lb	0.454
0.01	kilogram, kg	quintal (metric), q	100
1.10×10^{-3}	kilogram, kg	ton (2000 lb), ton	907
1.102	megagram, Mg (tonne)	ton (U.S.), ton	0.907
1.102	tonne, t	ton (U.S.), ton	0.907
Yield and Rate			
0.893	kilogram per hectare, kg ha^{-1}	pound per acre, lb acre^{-1}	1.12
7.77×10^{-2}	kilogram per cubic meter, kg m^{-3}	pound per bushel, lb bu^{-1}	12.87
1.49×10^{-2}	kilogram per hectare, kg ha^{-1}	bushel per acre, 60 lb	67.19
1.59×10^{-2}	kilogram per hectare, kg ha^{-1}	bushel per acre, 56 lb	62.71
1.86×10^{-2}	kilogram per hectare, kg ha^{-1}	bushel per acre, 48 lb	53.75
0.107	liter per hectare, L ha^{-1}	gallon per acre	9.35
893	tonnes per hectare, t ha^{-1}	pound per acre, lb acre^{-1}	1.12×10^{-3}
893	megagram per hectare, Mg ha^{-1}	pound per acre, lb acre^{-1}	1.12×10^{-3}
0.446	megagram per hectare, Mg ha^{-1}	ton (2000 lb) per acre, ton acre^{-1}	2.24
2.24	meter per second, m s^{-1}	mile per hour	0.447
Specific Surface			
10	square meter per kilogram, m^2 kg^{-1}	square centimeter per gram, cm^2 g^{-1}	0.1
1000	square meter per kilogram, m^2 kg^{-1}	square millimeter per gram, mm^2 g^{-1}	0.001
Pressure			
9.90	megapascal, MPa (10^6 Pa)	atmosphere	0.101
10	megapascal, MPa (10^6 Pa)	bar	0.1
2.09×10^{-2}	pascal, Pa	pound per square foot, lb ft^{-2}	47.9
1.45×10^{-4}	pascal, Pa	pound per square inch, lb in^{-2}	6.90×10^3

(continued on the next page)

Conversion Factors for SI and non-SI Units - (continued)

To convert Column 1 into Column 2, multiply by	Column 1 SI Unit	Column 2 non-SI Unit	To convert Column 2 into Column 1 multiply by
		Density	
1.00	megagram per cubic meter, Mg m^{-3}	gram per cubic centimeter, g cm^{-3}	1.00
		Temperature	
1.00 (K − 273)	Kelvin, K	Celsius, °C	1.00 (°C + 273)
(9/5 °C) + 32	Celsius, °C	Fahrenheit, °F	5/9 (°F − 32)
		Energy, Work, Quantity of Heat	
9.52 × 10^{-4}	joule, J	British thermal unit, Btu	1.05 × 10^3
0.239	joule, J	calorie, cal	4.19
10^7	joule, J	erg	10^{-7}
0.735	joule, J	foot-pound	1.36
2.387 × 10^{-5}	joule per square meter, J m^{-2}	calorie per square centimeter (langley)	4.19 × 10^4
10^5	newton, N	dyne	10^{-5}
1.43 × 10^{-3}	watt per square meter, W m^{-2}	calorie per square centimeter minute (irradiance), cal cm^{-2} min^{-1}	698
		Transpiration and Photosynthesis	
3.60 × 10^{-2}	milligram per square meter second, mg m^{-2} s^{-1}	gram per square decimeter hour, g dm^{-2} h^{-1}	27.8
5.56 × 10^{-3}	milligram (H_2O) per square meter second, mg m^{-2} s^{-1}	micromole (H_2O) per square centimeter second, μmol cm^{-2} s^{-1}	180
10^{-4}	milligram per square meter second, mg m^{-2} s^{-1}	milligram per square centimeter second, mg cm^{-2} s^{-1}	104
35.97	milligram per square meter second, mg m^{-2} s^{-1}	milligram per square decimeter hour, mg dm^{-2} h^{-1}	2.78 × 10^{-2}
		Plane Angle	
57.3	radian, rad	degrees (angle), °	1.75 × 10^{-2}

(continued on next page)

Conversion Factors for SI and non-SI Units - (continued)

To convert Column 1 into Column 2, multiply by	Column 1 SI Unit	Column 2 non-SI Unit	To convert Column 2 into Column 1 multiply by
Electrical Conductivity, Electricity, and Magnetism			
10	siemen per meter, S m⁻¹	millimho per centimeter, mmho cm⁻¹	0.1
10^4	tesla, T	gauss, G	10^{-4}
Water Measurement			
9.73×10^{-3}	cubic meter, m³	acre-inches, acre-in	102.8
9.81×10^{-3}	cubic meter per hour, m³ h⁻¹	cubic feet per second, ft³ s⁻¹	101.9
4.40	cubic meter per hour, m³ h⁻¹	U.S. gallons per minute, gal min⁻¹	0.227
8.11	hectare-meters, ha-m	acre-feet, acre-ft	0.123
97.28	hectare-meters, ha-m	acre-inches, acre-in	1.03×10^{-2}
8.1×10^{-2}	hectare-centimeters, ha-cm	acre-feet, acre-ft	12.33
Concentrations			
1	centimole per kilogram, cmol kg⁻¹ (ion exchange capacity)	milliequivalents per 100 grams, meq 100 g⁻¹	1
0.1	gram per kilogram, g kg⁻¹	percent, %	10
1	milligram per kilogram, mg kg⁻¹	parts per million, ppm	1
Radioactivity			
2.7×10^{-11}	bequerel, Bq	curie, Ci	3.7×10^{10}
2.7×10^{-2}	bequerel per kilogram, Bq kg⁻¹	picocurie per gram, pCi g⁻¹	37
100	gray, Gy (absorbed dose)	rad, rd	0.01
100	sievert, Sv (equivalent dose)	rem (roentgen equivalent man)	0.01
Plant Nutrient Conversion			
	Elemental	Oxide	
2.29	P	P_2O_5	0.437
1.20	K	K_2O	0.830
1.39	Ca	CaO	0.715
1.66	Mg	MgO	0.602

Glossary of Soil Science Terms

READER RESPONSE FORM

Errors or inadequate definitions:

_____
_____
_____
_____

Terms to add to the next edition (please supply your definition or a reference):

_____
_____
_____
_____

Errors in scientific or common names (please explain):

_____
_____
_____
_____

Scientific or common names to add to the next edition:

_____
_____
_____
_____

Other:

_____
_____
_____
_____

Your name and address:

_____
_____
_____
_____

Return to: S374 Glossary of Soil Science Terms Committee
               SSSA Headquarters
               677 S. Segoe Road
               Madison, WI 53711

or: gloss_soil@agronomy.org   Fax: 608-273-2021